Grove Karl Gilbert

AMERICAN LAND AND LIFE SERIES
Wayne Franklin, series editor

Grove Karl Gilbert

A Great Engine of Research

by Stephen J. Pyne

UNIVERSITY OF IOWA PRESS, IOWA CITY

University of Iowa Press, Iowa City 52242
www.uiowapress.org
Copyright © 1980 by the University of Texas Press
Author's Note copyright © 2007 by Stephen J. Pyne
Printed in the United States of America

Parts of this book are based upon the following articles
by Stephen J. Pyne: "Geophysics in the Giant Forest: G. K.
Gilbert as Conservationist," *Environmental Review*, no. 6
(1978); "Methodologies for Geology: G. K. Gilbert and T. C.
Chamberlin," *Isis* 69, no. 248 (1978); "Certain Allied Problems
in Mechanics: Grove Karl Gilbert at the Henry Mountains," in
Two Hundred Years of Geology in America, ed. Cecil Schneer
(Hanover, N.H.: University Press of New England).

The University of Iowa Press is a member of Green Press
Initiative and is committed to preserving natural resources.

Printed on acid-free paper

Library of Congress Cataloging-in-Publication Data
Pyne, Stephen J., 1949–
Grove Karl Gilbert: a great engine of research /
by Stephen J. Pyne.
p.   cm.—(American land and life series)
Originally published: Austin: University of Texas Press, c1980.
Includes bibliographical references and index.
ISBN-13: 978-1-58729-618-5 (pbk.)
ISBN-10: 1-58729-618-7 (pbk.)
1. Gilbert, Grove Karl, 1843–1918.   2. Geologists—
United States—Biography.   I. Title.
QE22.G5P96 2007   2007009518
550.92—dc22
[B]

07 08 09 10 11 P 5 4 3 2 1

Contents

Author's Note
to the Paperback Edition

It was a good bet that an additional 30 years would leave Grove Karl Gilbert still in fashion. It was a less sure thing that my biography of him might yet find an audience. That it has, I owe to Holly Carver, who has edited both productions; to the continued enthusiasms for earth science, which was consolidating a revolution when the text first went into print and has continued to probe new frontiers in the crust, the deep oceans, and extraterrestrial worlds; and, above all, to Gilbert himself, enduring, imperturbably sensible, the antithesis of the celebrity scientist or romantic culture hero. He was very much a man of his age, yet not bound by that era. He remains a favorite among practitioners, for all the right reasons. In 1979, to commemorate its centennial, the U.S. Geological Survey established an annual G. K. Gilbert Fellowship to reward its outstanding scientist.

What might I do differently were I to write the book today? Over the years some matters have sharpened, some blurred, and some amplified. Among thematic issues, I would emphasize Gilbert's prescient insights into lunar geomorphology and the reasons behind his reluctant misreading of Meteor Crater. The latter resulted from his standing as a man of his time. The belief then prevailed that meteors had metallic cores, of which Gilbert could find no relic at the site; the breakthrough came decades later with the identification of coesite as a mineral diagnostic of impact craters. His work on exogeology, however, aligned Gilbert with the future. As spacecraft have returned images of Venus and Mars and the new worlds among the moons of the outer planets, geomorphology has been asked to interpret their history. This study is, like Gilbert's of the Moon, one of surfaces, from which one must devise explanations of structure, process, and time. Gilbert was among the first—may well be the originator—of this style of geology, and deserves more credit than I gave him. A partial explanation is timing: the full text was written in 1976, a year before the *Voyager* mission launched, and in revised form went through production between 1979 and 1980, as the *Voyagers* first encountered Jupiter and Saturn.

Among regrets is a last-minute recommendation I accepted from the copyeditor to insert "a few GKs" to vary the litany of "Gilberts." The instinct was sound. I had relied on the fact that "Gilbert" has currency as both a first and last name to cover all my uses. Some variety was preferable, and since Gilbert signed his name "G. K.," the suggestion was reasonable. I should have stopped it. But not seeing it in print, I didn't appreciate the effect. If a variant was needed, it should have been "Karl," which is how he was known to his friends. The damage may be slight, but it was unnecessary. My obtuseness is partly explained by my conception of the book as an intellectual biography. I treated Gilbert's work as his life, viewing it through the literary prism in which I had been trained, New Criticism, with its obsession over close reading independent of biographical considerations. Personal details seemed less relevant, and were, in any event, sparse.

Had I the chance to do the biography over, I would also work more of the personality into the text. A few "Karls" would have helped, along with replacing a couple of landscape photos with photos of Karl on the land. I can add, further, that subsequent information suggests Fannie Gilbert did not suffer from prolonged coal-gas poisoning but from alcoholism. Such at least was the scuttlebutt around the Survey. It makes sense of her recurring convalescences and the unburdening Karl felt upon her death. As to his own death, I watched myself bonding to more than his mind as the story unfolded, and when his mortality was visible on the horizon, I found it necessary to interrupt the text and write down his final days before returning to the last chapter. The thought of climaxing his oeuvre only to have him die was disheartening. It was easier to write his closing years as a retrospective (as it were). That sentiment, too, I would have tried to infuse more generally into a revised text.

Some themes in this book I have returned to in subsequent studies. The Great Ages of Discovery concept, first announced by William Goetzmann, I have circled back to repeatedly, and have amplified in *The Ice*, *How the Canyon Became Grand*, and in several anthologized essays, most recently "Seeking Newer Worlds: An Historical Context for Space Exploration" in *Critical Issues in the History of Spaceflight*. I intend to elaborate the concept in future studies, perhaps using the idea as an organizing conceit for a general history of exploration. The Powell Survey, too, and particularly Clarence Dutton, reappear in *How the Canyon Became Grand*. And

I was able to midwife a new edition of Dutton's *Tertiary History of the Grand Cañon District* that included an interpretive essay as a foreword. Gilbert was a wonderful guide to that bunch and the bold lands they revealed.

Grove Karl Gilbert was my first book. It seemed miraculous that, having read books for so many years, I was able to write one. Now, as my twentieth book goes into production, that process has largely demystified, although where the ideas and words come from remains as ineffable as ever. *Gilbert* showed me it was all possible—possible to conceive and write, possible to publish, possible to go on to another project. He was a good mentor: a great mind, a kindly man, a person who could reconcile field and study, a prodigious scholar. I hope my biography of him, or at least of his career, suffused as it is with the flaws and vitality of my youth, returns a portion of that generous example.

Steve Pyne
February 2007

Grove Karl Gilbert, 1898. *Courtesy of the USGS Photographic Library.*

Preface and Acknowledgments

Edwin McKee tells the story of the U.S. Geological Survey's Colorado River Expedition during which, over an evening campfire, the participants debated the question of who was America's greatest geologist. They selected Grove Karl Gilbert.

The setting as much as the choice is appropriate. G. K. Gilbert was not only a superb scientist—the only man, for example, to be twice elected president of the Geological Society of America—but he was an explorer of international renown. He continues to be remembered as much for his contributions to the scientific discovery of the West as for the unexcelled monographs, experiments, and statements on method which helped make American geology supreme in the early twentieth century. What happened around that campfire in the gorge of the Grand Canyon has been repeated dozens of times, in field and office alike. Gilbert, it is affirmed, was a real man of genius in a discipline known more often for its adventurers than for its intellectuals. Within the earth sciences, his reputation flourishes: his name is constantly invoked in favor of a method, a theory, or a precedent-setting investigation.

His biography deserves to be modernized. More than fifty years ago, William Morris Davis published a memoir of Gilbert, the largest in the National Academy of Sciences' biographical series. Although Davis evidently felt that no future biography would be necessary, he neglected to document his sources, and much of his research materials have eroded into oblivion. He also labored mightily to incorporate Gilbert into the theoretical superstructures of the age's geology, in particular, to place Gilbert into the context of Davisian geomorphology. Yet that endeavor, however sincere, introduced a terrific dissonance to the biography, and Davis' geology is now so dated that what he intended as flattery has become mere incongruity. It would be difficult to find two geologists more at odds in their perception of the earth.

Gilbert's reputation currently enjoys a renaissance of sorts. His mechanical analogies harmonize well with the increasing mathematicization of the earth sciences. His perspective on geologic time and on the analysis of geophysical and geomorphic systems resonates easily with the concept of a steady state and the appeal to systems theory. His papers on method are continually rediscovered and advertised. So thorough were some of his field studies and experimental researches that no one has attempted to reexamine them. The transformation of geologic thought which has antiquated the summae of his contemporaries has only revitalized appreciation for his own contributions. It is entirely appropriate that the Geological Society of America, which he helped found, should sponsor a symposium on his career as a means of celebrating the centennial of the U.S. Geological Survey—that "great engine of research," as Gilbert termed it—which he helped make into a world-famous institution.

The essence of this transformation in geologic thinking has been in the appreciation and organization of geologic time. A general systems or cybernetic interpretation of earth systems has replaced the evolutionism which saturated the earth sciences of Gilbert's age. In the nineteenth century, as information about the earth accumulated, people became aware of an enormous past. History emerged, first, to causally unite this new information and, finally, to provide a convenient nonetiological framework by which to order it. It was natural enough that evolution should provide the fundamental patterns for organizing and interpreting earth history. In the twentieth century, however, as more information has poured in, that framework has collapsed. The forces of the present, especially the human agency, have become overwhelming. The scale of effective geologic time and space has been abbreviated; time's arrow has been given a feedback loop; the steady state of an open system has replaced the directionality of processes in a closed system acting under evolutionary or thermodynamic considerations. New chronometers have emerged, like the rhythms of paleomagnetic reversals, to replace the linear chronologies of the fossil record and radioactive decay. Where the nineteenth century would have looked at the growth and metabolism of an organism, the twentieth prefers its homeostasis. Where the nineteenth might have looked to the tree as an emblem of nature's organization, the twentieth would use a brain or a computer.

The nineteenth century built up a complicated design of history, both natural and human; the twentieth has largely broken that pattern. Yet Gilbert was less an anticipator of the modern assault on historicism, founded on such concepts as simultaneity, complementarity, and indeterminacy, than he was a throwback to the mechanics of Newton. What is perhaps most remarkable about his achievement is that he continues to have vitality for a rapidly changing science. It is doubtful that this biography will be his last. But, by trying to show how Gilbert's education, temperament, perception, modes of expression, and science all intersect, I hope that a portrait of the man will emerge which has internal consistency, validity, and staying power. Between the idea of the man and the character of the man there seems to exist a kind of indeterminacy. I hope that my analysis of Gilbert's scientific works and of the systematic contrasts he makes when compared biographically and intellectually with his contemporaries will result in a workable compromise—fair equally to Gilbert, his age, and the present. It is only in this way that a biographer can hope to overcome to some extent his own historical circumstances.

Clarence Dutton once wrote of his association with Gilbert and John Wesley Powell that "the extent of my indebtedness to them I do not know. Neither do they. I only know that it is enormous, and if a full liquidation were demanded, it would bring me to bankruptcy." I can easily apply that statement to my own situation. For the Colorado Plateau, substitute this biography; for Gilbert and Powell, insert William Goetzmann, John Sunder, Victor Baker, and Alexander Vucinich. So much of this book is a response to their insights and questions that I cannot hope to measure their individual contributions. I can add with respect to Goetzmann's aid, however, a remark Gilbert made of Powell—that the least number of his ideas were those he published. The far greater number were those he circulated among his associates. In conversations and seminars I have been the beneficiary of the force field of Goetzmann's ideas. That I should compare him in this way to Powell is entirely suitable; I can think of no more fitting, or more deserved, compliment. To Jack Sunder, I owe most of my thinking about the West and, indirectly, about the conservation of the western landscape. To Vic Baker, I can trace volumes of my thought about the philosophy and concepts of modern geology. As for Alex Vucinich, he already knows that his considerable intellectual guidance in the history of science has been the very least of his help.

So many people and institutions furthered this project that I can name only a few. Three, in particular, helped with the accumulation of primary materials as well as with encouragement. I must thank, first of all, the late Karl Gilbert Palmer, grandson of G. K. Gilbert; his wife, Dorothy; and their two daughters for letting me examine their collection of documents. In this regard I must also acknowledge my debt to Arvid Johnson, L. H. Lattman, David Pollard, and Donald McIntyre for helping me track down the Palmers. Second, Donald Coates of SUNY Binghamton complemented the Palmer materials with a cache of Gilbert letters he kindly let me examine. Finally, the Smithsonian Institution funded a considerable portion of my research with a ten-week grant. I was extremely fortunate to work under the direction of Nathan Reingold, editor of the Joseph Henry Papers, and one of his assistant editors, Michele Aldrich. They made my sojourn to Washington vastly more pleasant and efficient. A fellowship to the National Humanities Center assisted in the final stages of manuscript preparation.

I would also like to thank many others who helped at particular points of the research and writing: Ellis Yochelson, for his written advice on research topics; Clifford Nelson, for pointing out several overlooked items of Gilbertiana; Richard Mahard of Denison University, who escorted me to many of the books Gilbert had donated to the school library; Edwin McKee, for pointing out materials in the U.S. Geological Survey Field Records files; John Hack, for conversations about Gilbert and geomorphology; Robert Crunden, who read and criticized the manuscript; Ronald DeFord, who assisted with research materials at the start of the project; my brother, Jim, who assisted at several technical points; the immensely helpful archivists at the National Archives and Smithsonian Archives, who guided me through their wilderness of documents; Holly Carver, whose editorial skills vastly improved the original manuscript; and my wife, Sonja, who contributed moral support and editorial commentary.

Grove Karl Gilbert

1. In a Nutshell

In later years, when his residence was elsewhere, visits to Rochester were frequent; he nearly always halted there on his journeys to the West and back. When returning from the Henry Mountains of Utah in the autumn of 1876 he was in time to attend his parents' golden wedding on November 30.

—William Morris Davis

Grove Karl Gilbert was born May 6, 1843—ten years after Charles Lyell published the final volume of his *Principles of Geology;* he died May 1, 1918—ten years before the proceedings of the first international symposium on continental drift were published. The seventy-five years of his life consequently spanned the heroic age of American geology, the period during which the science was intellectually and institutionally defined. Gilbert knew most of its grand figures—James Hall, James Dwight Dana, John Strong Newberry, John Wesley Powell, Clarence Dutton, Joseph Le Conte, Thomas Chrowder Chamberlin, and William Morris Davis. And he himself contributed impressively to the geology of the heroic age in its several cultural functions.

Geology evolved as a mechanism for coping with certain intellectual problems, particularly the consciousness of a landscape whose spatial and temporal scales were rapidly expanding. Prior to the latter half of the eighteenth century, western civilization's geographic knowledge was limited to the coastlines of most of the world's continents. The only continental interior known with any precision was that of Europe. Equally scant was the recognized span of time. History began, according to Bishop Ussher's genealogical calculations, in 4004 B.C. Before Gilbert's birth, these narrow scales of time and place were shattered, as European and American explorers busily unveiled the interiors of all the continents and as evidence of past civilizations, organisms, mountains, and continents was discovered in the form of ruins and fossils. It was from the ensuing debate about the age and size of the earth that geology emerged. In particular, the unspeakable vastness of time became its special province. The resolution to this debate, which spanned nearly 150 years, came during Gilbert's lifetime. In a general way, the debate determined the shape of his career. He himself became one of the premier scientific explorers of the American West, thus contributing to the cascade of new information about the earth. At the same time, he addressed the theoretical questions which sought to interpret this new data and assimilate it into scientific form. His answers to the larger questions of the earth, as well as the methodology he proposed to accompany them, are classics of the science and continue to have significance.

For America, moreover, geology had additional importance as a frontier institution, an economic and intellectual subsidy to the westward migration. The discovery of landforms, rich new soils, lodes of precious minerals, water resources, breathtaking vistas, and scenes of high adventure coincided with an outburst of cultural nationalism and a sprawling folk migration across North America. The developing sciences of the earth could not only help uncover that landscape but could aid in its assimilation as well; they could delineate prime sites for agricultural settlement, for industry, and for aesthetic appreciation. In short, the geologic sciences helped incorporate the western landscape into American political, social, and economic institutions as much as into its intellectual heritage. That Gilbert should spend the most profitable years of his career in the Far West was only natural: the opportunities were greatest there. At the same time, his work repaid those opportunities by skillfully pursuing problems of significance to western settlement and to the conservation of the western landscape.

By the time Gilbert's career ended, American earth science had entered into the mainstream of American culture and ranked as one of the outstanding national traditions in science. It had a broadly based institutional foundation, enormous reserves of data, and a theoretical superstructure which not only answered the debate which first spawned geology but successfully integrated those concerns with the physical, biological, chemical, and social sciences. Equally significant, American geology had a magnificent tradition— almost a mythology—of exploration and insight which grew out of the experiences of the heroic age. So powerful was this tradition that, until very recent times, the story of American geology rightly meant the story of its heroic age. The biography of Grove Karl Gilbert belongs in the chronicle of that tradition, not merely as an especially curious episode within it but as the narrative of one of the tradition's founders.

The Education of a Classicist

One of the surprising aspects of Gilbert's career is the fact that he ever became a geologist at all. In some respects improbable, his scientific calling can nevertheless be triangulated from two bench marks of his youth: his family environment and his education. The progenitor of the American Gilberts arrived in Massachusetts in 1630, but, if we accept a joking remark by G. K. Gilbert, we might thrust the family line further back. Thanking a paleontologist friend for naming yet another fossil after him—"I see you have given me a new ancestor," he wrote William Dall, "and a very good looking one"—Gilbert threatened to construct a "paleogenealogy" and to "gloat over the DARs with their brief historical pedigrees." The significant genealogical fact, however, is not the longevity of the family but the unorthodox behavior of one of its members, Grove Sheldon Gilbert, GK's father. Breaking with family tradition, he left New England for Niagara, Toronto, and finally Rochester, New York; he abandoned the trades of cooper and machinist to become a self-taught portrait painter; and he gave up the religious orthodoxy of Presbyterianism for speculations more to his liking. Although Grove Sheldon reportedly possessed a "deeply religious nature," no formal instruction in religious matters devolved on his children. Grove Sheldon's apostasy, an act of secularization, occurred about the time Grove Karl Gilbert was born.[1]

The Gilbert family was a nearly self-contained environment. Warm but private, relying heavily on its internal resources, it was affectionately and appropriately referred to as the Nutshell. Frugality was a necessity; entertainment consisted largely of the "intellectual game," as GK called it, in which evenings were spent with riddles, round-robin poetry composition, spelling contests, card games, and reading aloud from books. It was a world which revolved around the dynamic, opinionated, unorthodox father—little is known of Gilbert's mother, Eliza—and it became the archetype for all Gilbert's later associations. Throughout his life, he tried to recreate it whenever possible, usually finding it in scientific clubs, his own homelife, and frequent visits to his older sister, Emma, who moved to Michigan, and his older brother, Roy, who remained at the Nutshell. Characteristically, G. K. Gilbert gravitated toward men who resembled his father—men like John Newberry, John Wesley

The West of G. K. Gilbert

Powell, and C. Hart Merriam. With the addition of billiards, his preferred entertainments were variants of the "intellectual game."[2]

Gilbert's education was equally informative. Outside of the Nutshell, he had little social life; he "studied a great deal when not working," as he put it. His studiousness rapidly became a hallmark. His health was too erratic for him to attend the local schools regularly, so the family assumed the chore of educating him. In astonishment at his tenacity, his aunt once exclaimed that his "absorption of tasks is like the greed of the horse leech—give, give is his eternal cry. I believe he could learn double the number of lessons if required." When he was able to attend the public schools, the report was the same; one schoolmaster, though with a stammer of pedantry and perhaps with an eye to his continued employment, wrote that young Gilbert was "*a most faithful, industrious, and attentive student, meriting in every way my highest approbation.*" While the schoolmaster's imagery was less vivid than the aunt's, the observation was virtually identical. When formal assignments lagged, Gilbert read extensively in the penny magazines.[3]

Clearly his father exerted a special force. The artistic abilities which brought Grove Sheldon an honorary membership in the National Academy of Design in 1848 he taught to his son. The awed aunt thought the younger Gilbert suffered from want of method, chauvinistically interpreting this as a mark of genius, but the transfer was incomplete. Except for his draftsmanship on the Ohio Survey, Gilbert's drawings never went beyond the landscapes sketched in his field notebooks and the portraits of mules he doodled while in the Utah desert. He quickly turned to photography instead. More significant was that special education his father transmitted in the form of riddles. These must have been a source of enjoyment as well as instruction, for riddles became habitual with him. Companions of his mature years often recalled how, during breaks in fieldwork or around an evening campfire, Gilbert challenged them with problems no doubt similar to those his father had posed.[4]

But the "horse leech" curiosity in him craved more, so his family sent him—alone among their children—to college. In 1858 he matriculated at the University of Rochester, though, as a means of insuring the substantial family investment, with the stipulation that he shore up his health with a program of outdoor exercise. Exploration geology became a fulcrum for maintaining that educational proportion throughout his adult life; his hours of boating on the Genesee River paid off later amid the rapids of the Grand Canyon and the fjords of Alaska. At the university he received a traditional, classical education. Biased toward languages, the prescribed program included rhetoric, logic, French, German, Greek, and Latin. How much of the languages stayed with him is debatable. Certainly, outside of coining an occasional scientific term from Greek roots, he had little use for the ancient languages. The modern languages he probably soon forgot, if indeed he ever learned them beyond the elementary stages. In later life he relied on translators to converse in French during an 1888 trip to Paris, and he was similarly dependent on translation for readings in the literature of French science. Davis suggests that Gilbert once tried to revive his study of French and German but without much success. Except for translated texts, Gilbert remained relatively ignorant of European science other than British. Merriam, on the other hand, recalled that "the few books he kept close at hand were mainly Greek, Latin, French, German, and English dictionaries; British and American encyclopedias, and technical works on geology, astronomy, and mathematics." Of that set the science books were undoubtedly the most heavily used.

The volumes in his reference library represented well his science curriculum at the university. Except for a few courses in geology and botany taught by Henry Ward, the sciences were of a classical variety—a blend of physics, astronomy, and especially mathematics collectively known as natural philosophy. At this time physics was consolidating the newly defined laws of thermodynamics, but, as John Merz has observed, the chief discoverers tried earnestly to bring the new "energetics" into "harmony and continuity with the older Newtonian laws. . . . The *Principia* of Newton was again studied, and re-edited in the unabridged form, and an interpretation and amplification of the third law of Motion—so as to embrace the principle of energy—was made the key to the science of mechanics." It was in this context, probably through the medium of the *Manual of Applied Mechanics*, by William Rankine, the Scottish physicist and engineer, that Gilbert learned the fundamental sciences.[5]

When he graduated at age nineteen, his largest block of credits was in mathematics, but he carried away an award for his Greek, the presidency of one of the two literary societies on campus, and an indelible bias toward classicism which stamped his subsequent intellectual career. That impression appears in his prose style, his scientific concepts, his philosophy of method, and his metaphysical assumptions about nature. It merged readily with the scholarly temperament of a man who was already reserved, imperturbable, serious, quietly unconventional, famous for his self-control and his often self-deprecating humor.

By all evidence that temperament formed early and, like the

A clerk in the cosmos. Gilbert in 1862 at age nineteen, the graduate from the University of Rochester. *Courtesy of the National Academy of Sciences Archives.*

crisp landscapes he later analyzed, it preserved its shape through time. The stories about GK's childhood are few. That in itself is telling, for he rarely reminisced except to illustrate some point of scientific observation. Those stories gleaned from associates may be apocryphal but, if they err on the child, they hold true for the man. There is Gilbert the superlative observer, discovering onion sprouts in the family garden before anyone else could see them. There is Gilbert, the man who loathed controversy and personal confrontation, being shoved from behind down the school stairs but not wishing to discover the culprit for fear he "should not like him anymore." There is Gilbert in his dignified shyness, the youth who fell through the ice while skating on Irondequoit Bay, dried out his clothes at a nearby house, and told no one when he quietly returned home. And there is Gilbert the amateur inventor, the warm ironist, the thorough mechanist as revealed in the story of the *Great Western*. A race on the Genesee was scheduled to celebrate the launching of Brunel's *Great Eastern* steamship, a colossal piece of heroic engineering. Gilbert created a featherlight craft of unconventional design involving experiments in mechanics whose trial runs frightened off his competitors. With a splendid example of his characteristic mock heroism when describing his own efforts, he christened his ship the *Great Western*.

Perhaps one can best epitomize Gilbert's education by recounting an incident involving one of the innumerable riddles his father confronted him with. Gilbert continued to recall it many decades later. Grove Sheldon posed the question this way: "A loaf of bread is in the shape of a hemisphere, the volume of the crumb equaling the volume of the crust. What are the dimensions of crumb and crust?" It is told that when Gilbert returned the correct answer his father proclaimed that, despite the boy's uncertain health, he might amount to something after all.[6] Actually, Grove Sheldon was too reserved in his judgment. Those childhood riddles became the model for Gilbert's gigantic monographs; inadvertently perhaps, they communicated a philosophy of science. Grove Karl Gilbert had reason to recall that particular problem of the bread loaf, for in 1877, with the publication of his *Report on the Geology of the Henry Mountains*, he solved a question in geophysics almost identical in form. His solution, perhaps drawn from his subconscious memory of childhood experience with crust and crumb, made his international reputation.

A Clerk in the Cosmos

In 1862, however, despite his baccalaureate and indifferent to the Civil War which raged to the south, the nineteen-year-old Gilbert had not advanced noticeably toward a formal career. His college degree was a source of satisfaction but also of debts. Part of its cost came in the long hours of study, part in the financial sacrifice by his family, and part in a social isolation that Gilbert was probably already inured against. His round of amusements had largely confined itself to the Nutshell; his apparel was homely and often badly worn. It was probably during these years that he learned to mend his own clothes—a habit which stayed with him permanently. A preference for homespun, a talent for personal frugality leavened by generosity toward others, and a restricted round of amusements—the

habits of these early years define the circumference of the social circle of his later years.

But they did not bring Gilbert a job, and they did not erase his
nagging debts. War and crusading ideologies had little appeal to a
man of Gilbert's reserve, so he never enlisted in the Union army,
and, probably due to his uncertain health, he wasn't drafted. Instead,
unfortunately, he tried to capitalize upon the one job his education
had fitted him for—education—and he journeyed to Jackson, Michigan, to initiate the experiment. On the surface the job seemed a
workable solution; in actuality it omitted one crucial ingredient:
GK was temperamentally unsuited to teach. He might have coped
with an urban or a college crowd but not with rural rowdies. He
simply could not control his students. As an apprentice school-
teacher or an established scholar, he abhorred the rod. Before 1862
ended he resigned. In another sense, the episode demonstrated the
intellectual personality of a man who, over twenty years later, reflected that education was a process not of instructing by rote but
of teaching by example. The example of a frail nineteen-year-old
scholar of mathematics and classics was not apt to impress the
youths of frontier Michigan, and Gilbert retired in ignominy.

Here for all its apparent indecisiveness was a pivotal moment.
There were no job prospects, and, perhaps equally critical, there was
no apparent mechanism for leaving the comfortable environment of
the Nutshell. Gilbert called this period of his life demoralizing. "I
had no heart to do the various things I had supposed I very much
wanted to when I was too busy to find time," he remembered. "Wait-
ing for something to turn up seems to be an occupation in itself."
Fortunately, he was no better at this occupation than at schoolteach-
ing. By 1863 he landed his first permanent job, an apprenticeship at
Cosmos Hall—a scientific factory for natural history organized by
Henry Ward on the Rochester campus.[7]

This was a marvelous compromise, a continued moratorium in
the form of a graduate school. The establishment was an institutional expression of the explorations which had scoured the globe
on behalf of western civilization for nearly a century; in America it
corresponded to the grand reconnaissance of the West begun by
Lewis and Clark and expanded by the Wilkes Expedition and the romantic surveys of the Army Corps of Topographical Engineers. It
was a clearinghouse, a scientific storeroom for processing the crush-
ing onslaught of natural history specimens which made Ward the industrial counterpart to the academician Agassiz, the adventurer Frémont, and the artist Audubon. But the establishment took its name,
as well as its inspiration, from Alexander von Humboldt's summa of

natural history, *Cosmos.* Cosmos Hall was an intellectual as well as an economic machine for processing the artifacts of exploration and then distributing them, usually to museums, where they could help educate the public.

Gilbert labored as a general clerk, counting, sorting, identifying, and labeling thousands of specimens—work he later euphemistically termed practical experience. For five years he accepted this tedium philosophically, and, while not overcome by a Humboldtean fever at the sight of mountains of unnamed specimens, he later remarked graciously that Cosmos Hall "thus served incidentally as a training school in the natural sciences and especially in certain branches connected with museums." At the same time, it genuinely encouraged his education by allowing him to continue his studies in the evening, usually in mathematics, occasionally in anatomy and geology. If he did not have it earlier, Gilbert's years at Cosmos Hall crystallized a powerful distrust of taxonomies. Moreover, his tenure at Ward's scientific establishment allowed him to gradually widen the scope of his associations. Some of his fellow workers, such as E. E. Howell, reappeared in his later career; others, such as James Hall and John Newberry, were introduced to him through Ward, and under them Gilbert respectively published his first scientific paper and became a journeyman explorer.[8]

Many years later, Gilbert properly listed himself "somewhat proudly," if generously, among the alumni of Cosmos Hall. However monotonous its discipline or anachronistic its purpose, the establishment spared him for science. Through its office routine and occasional field expeditions, it initiated him into a long apprenticeship in geology. It was, in fact, on one such assignment—to help excavate and reconstruct a fossil mastodon near Albany—that Gilbert's scholarly incubation ended and he became committed to geology. Institutionally, Ward's establishment provided a niche in an age not overwhelmed with scientific jobs but, philosophically and temperamentally, it left Gilbert untouched. He could work in the Rochester of Cosmos Hall, but it was a long leap from the tidy world of the Nutshell to the panoramic drama of the Humboldtean cosmos, one Gilbert never attempted.

Yet it is only in the context of this chasm that Gilbert's geologic contributions can be understood. Alexander von Humboldt was the most glamorous and articulate exponent of the age of exploration which, among others, sent Cook, Bougainville, and Wilkes to the Pacific, the Institut de France to Egypt, Pallas to Russia, Bering to Alaska and scattered a host of explorers and adventurers into the interior of the world's continents. More than anyone else, Humboldt defined what that new reconnaissance meant. When he convinced the Spanish monarchy to let him tour South America for five years, from 1799 to 1804, he was called a second Columbus, and his *Voyage to the Equinoctial Regions of the New World* created an appreciation and an understanding on whose model other newly discovered or rediscovered continents—Australia, Antarctica, Africa, Asia, and North America—entered European consciousness. Humboldt's *Personal Narrative* established the vogues of the traveler, the mountaineer, and the scientific discoverer of the remote and sublime. His Thomistic, multivolumed *Cosmos* attempted to summarize, in vivid and popular language, both the new impression and the expanded knowledge of nature freighted back by the explorers. For his labors, Humboldt became the most visible scientist of his age: he epitomized the naturalist as hero. What Beethoven was to the music of the romantic period, what Napoleon was to its politics, Humboldt was to its science.[9]

He exercised special appeal to Americans. An ardent democrat, he expressed great hopes for the young republic, and the young republic responded enthusiastically. Emerson called him "a universal man," . . . "one of those wonders of the world, like Aristotle." Whitman, the self-styled "poet of the cosmos," attempted to synthesize with verse ensembles what Humboldt did with geographic science. Poe dedicated his metaphysical prose poem "Eureka" to this giant of scientific adventure. John Lloyd Stephens traveled to Yucatán to rediscover the American Egypt which Humboldt had popularized, and, at the age of seventy-six, John Muir realized a lifelong ambition when he journeyed to the headwaters of the Amazon to find the flowers and trees Humboldt had described.

Humboldt was the ideal exemplar for a nation pushing into a western wilderness. The chief explorers of the American West prior to the Civil War—the Army Corps of Topographical Engineers—took him for a polestar; and on his example dozens of German scientists swarmed over the West, either on private surveys or, attached in a technical capacity, on American expeditions. In the same way Humboldt inspired a whole range of American academicians, from Frémont the explorer to Agassiz the professor of natural history at Harvard. Although a student of Cuvier's, Agassiz imagined himself as a personal emissary of Humboldt's, who several times befriended him as a patron. Like Humboldt, Agassiz planned a total inventory of North American natural history, modeled on *Cosmos*. Appropriately enough, when Humboldt visited the United States in 1804, he was entertained by Jefferson the same month that

Lewis and Clark left St. Louis for the Rocky Mountains and the Pacific.

The comparison of Humboldt to Columbus, frequently made, is apropos: he not only uncovered a new world for science but publicized a new field of scientific study, natural history. Actually, he had not been the first European scientist in South America—La Condamine had preceded him in 1735—but he was the first to systematically reveal its flora, fauna, climate, rocks, ruins, and native peoples, in short, its geography. Humboldt's survey dealt with the life and earth sciences, La Condamine's with questions in Newtonian physics. For five privately financed years, accompanied by Aimé Bonpland, Humboldt traveled through Venezuela, Colombia, Ecuador, Peru, Mexico, and Cuba; he clambered up mountains, singeing his boots on volcanoes and setting a world altitude record on the flanks of Mount Chimborazo; he paddled up the Orinoco River to the inconspicuous channel it shares with the Amazon; and everywhere he measured and mapped, collecting specimens by the thousands. His was a passionate quest, at times almost demonic, as a letter from Venezuela dramatizes:

> What a fabulous and extravagant country we're in! Fantastic plants, electric eels, armadillos, monkeys, parrots: and many, many, real, half-savage Indians . . . Up till now we've been running around like a couple of mad things; for the first three days we couldn't settle to anything; we'd find one thing, only to abandon it for the next. Bonpland keeps telling me he'll go out of his mind if the wonders don't cease soon.[10]

But the wonders did not cease. With scientific genius and Promethean flamboyance, Humboldt proposed a scheme to absorb the new data and visual impressions. "I have the crazy notion," he wrote a friend of his ambitious design for *Cosmos*, "to depict in a single work the entire material universe, all that we know of the phenomena of heaven and earth, from the nebulae of stars to the geography of mosses and granite rocks—and in a vivid style that will stimulate and elicit feelings."[11]

The critical word in this passage is "geography." Humboldt pioneered in the systematic collection and integration of geographic data—with concepts of regionalism, plant geography, isothermals, the geologic cross section, political economics, ocean currents, and lines of equal intensity of terrestrial magnetism. In addition, he extended geographic concepts to investigations in archaeology and anthropology. His was an improbable blend of the German *Naturphilosophie* and the French *Encyclopédie*. Humboldt tried to integrate the thousands of newly discovered species and strata through physical laws of geography, organizing them into ensembles of plants, suites of rocks, and communities of human settlement. In doing so, he mathematicized his novel information by hurling it onto a sort of Cartesian coordinate system, the map, then trying to relate the burgeoning individual points through equations of space. At the same time, he helped give a new aesthetic and cultural meaning to novelty itself: it was not intrinsically threatening but sublime.[12]

The traditional framework for cataloging the inhabitants of the earth was the great Chain of Being. Responding to the steadily increasing quantities of specimens, Linnaeus had multiplied the single chain into several for the organic world. On his example, Abraham Werner had done the same for the inorganic world. What Humboldt, a Werner student, achieved was the addition of a second axis to the set of chains. The result was a map which showed the natural geographic assemblages of rocks and landscapes. Humboldt thus invented a comparative geology to match the comparative anatomy that Cuvier was developing in the same period. What the system lacked, however, was what Arthur Lovejoy has termed "the temporalization of the Chain," that is, the addition of a time axis. It was the conversion of the chain into historical links that brought about the great nineteenth-century syntheses in biology and geology. In short, Humboldt tried to organize an increasingly complex universe through laws of space, while most thinkers followed Hegel's attempt to use laws of time. Consequently, Humboldt's thought was more a continuation of Diderot than an anticipation of Darwin. The result rendered *Cosmos* into the grand summary and gesture of a romantic deist.

It also made it obsolete almost as soon as it was published, a fact made vivid by noting that in 1859 Darwin published *On the Origin of Species*, a few months after Humboldt's death at age ninety. And where Humboldt, by rejecting history as an organizing principle, also ignored the core enigma of geology—the fossil—Darwin made it integral to his system of evolution. While Humboldt successfully resolved one aspect of the new information, an expanded knowledge of geographic places, Darwin answered the other aspect, an enlarged understanding of nature's history. Thus, while geography addressed the problem of the earth's size, it was paleontology which was destined to resolve the central debate of geologic science through the century: the age of the earth.

This was a debate Humboldt largely ignored. Consequently, long before Darwin and Alfred Wallace worked out the mechanism of organic evolution, and before Kelvin and Helmholtz hypothesized the heat history of the earth and sun, Humboldt was already a splendid anachronism, an emblem of the rapid accumulation of information rather than of its intellectual assimilation—as the young clerk labeling specimens in Cosmos Hall must have been painfully aware. Yet Humboldt did furnish the basic handbook for the new explorations, and it was for his example as the quintessential explorer that he was celebrated, for his adventurous *Personal Narrative* rather than his encyclopedic *Cosmos.*

As long as there remained the challenge of uncharted jungles, windswept deserts, and the icy stillness of the poles, however, the Humboldtean tradition had meaning. It supplied the fundamental pattern for the exploration of European colonial and scientific frontiers, not only in North America but in India, Australia, Africa, and even China. It sent Gilbert on the Wheeler, Powell, and Harriman expeditions to the Grand Canyon, Death Valley, the Henry Mountains, and Alaska. It culminated in Gilbert's lifetime with the publication of that monumental encyclopedia of the globe, Eduard Suess' *The Face of the Earth.* And, even though as an intellectual enterprise it was eclipsed by Darwinian evolution and thermodynamics, Darwin himself generously acknowledged the debt he owed the grand explorer of the cosmos. When he departed on the *Beagle,* as another participant in that panoramic era of exploration, he packed three books with him: the Bible, Lyell's *Principles,* and Humboldt's *Personal Narrative.* They were three books that American geologists continued to carry throughout the century.

Of Mastodons and Mathematics: The Cohoes Potholes

All that romantic splendor was as distant from the Nutshell as the Orinoco was from Rochester. Of the three representative books that Darwin carried around the world, Gilbert absorbed only Lyell with any consequence. He read the Bible, if he read it at all, as a piece of secular literature, in the same spirit that he attended sermons—as though they were lectures at the New York Lyceum of Natural History. His language and imagery were utterly free of biblical allusions. Though an explorer, he never shared the Humboldtean sense of awe and drama; though a geologist, he integrated the processes of the geological environment in ways apart from those of either Humboldt or Darwin. The fossil he systematically ignored. His mathematics was not of cartography. His elementary laws of nature were neither historical nor geographical but physical—expressions of Newtonian mechanics. The sheer dimensions of the Humboldtean explorer were emotionally foreign to the condensed world of the Nutshell. Combined with his basically classical education, GK's entry into geology is, therefore, somewhat surprising. The point of conversion came inadvertently in 1866 when, on assignment for Cosmos Hall, he journeyed to Cohoes Falls near Albany.

Amid a channel of gravel-filled potholes, James Hall was supervising the excavation of a mastodon. The discovery of a mastodon could be a major cultural event, akin to finding the ruins of a lost civilization—from their status as an enigma, the enormous skeletons had become a crucial component of natural history. Just as the discoveries of exotic places added to a romantic geography, these ancient relics evoked the image of a romantic history. And, ever since Jefferson had defended the splendor of the American environment against the snobbery of Buffon by citing its prodigious mastodon finds, the mastodon had acquired special significance as an emblem of emergent nationalism. Though in ways different from Humboldt's ascent of Chimborazo and Darwin's exploration of the Galápagos archipelago, the excavation of these fossil giants also signified the discovery of natural history. Salvaging a few such wondrous relics, for example, could almost redeem a state geological survey from its otherwise fatal inability to locate gold or coal.

When Hall fell into the pit and injured himself, Gilbert took over—a routine project suddenly turned into a productive and educational stroke of luck. In the course of actually reconstructing the skeleton, Gilbert, accompanied by E. E. Howell, visited two assembled skeletons in Boston, met such scientific luminaries as Louis Agassiz and Jeffries Wyman, and broadened his circle of acquaintants to include workers at the Albany and Columbia University museums. In March 1867, he published a newspaper article on the project in Moore's *Rural New Yorker*. In all, he did a creditable job and, for his meticulous labor, was rewarded by later assignments with Irish elks.[13]

After he completed the reconstruction of the Cohoes mastodon, Gilbert returned to the site to examine something that intrigued him far more than the ancient bones: he studied the potholes. In fact, he surveyed all of them in the vicinity, counted them (350), measured them, and described their shape (like a "chemist's test tube"). He proposed a cause of origin: they were produced by "the grinding action of stones moved by water." Dangling by ropes to where he could cut trees and count growth rings, he estimated the rate of recession for the falls: twelve inches a century, or approximately 35,000 years as a "minimum for the time that has elapsed since Cohoes Falls were opposite the Mastodon pothole." As for the falls itself, he concluded that it was probably "the normal mode of descent of a river over these upturned shales." That is, rather than reconstruct the evolution of the mastodon, Gilbert preferred to reconstruct its physical environment. He tried to convert a problem of geography and history into one of geometry and physics. In his newspaper account he had considered the bones as a structural element, an expression of a physical law relating bone size to strength; in his pothole essay he similarly tried to discover a mathematical ratio for a physical structure, an equation for a natural process.[14]

In the course of this investigation, Gilbert became seriously attracted to geology. Recalling his graduation from the University of Rochester, he later remarked that, having been exposed to both, at the time he had found engineering more interesting than geology. Amid the Cohoes potholes, however, he realized happily that the two could be combined. It was his special genius to apply methods commonly used in hydraulic and civil engineering to geologic subjects.

The study of the Cohoes potholes thus prefaces all of Gilbert's mature work: it addressed problems posed by the environment of upstate New York; it showed a reserved classicist responding to discoveries that made ardent romantics out of most of his associates; and it turned fossils into physics, transforming an artifact of the past into a mechanism of the present. It also, not incidentally, sparked a lifelong fascination with potholes, which he pursued with equal enthusiasm in the gorge of the Grand Canyon and on the slopes of the High Sierras. He even imagined Niagara Falls—the successor to the Cohoes study—as a pothole turned on its side.

When Gilbert termed Niagara a physiographic engine, when he conceived of the Henry Mountains as a gigantic piston, when he gave an address "On the Uses of the Canyons of the Colorado for Weighing the Earth," he was elaborating on the methodological lessons learned while investigating the Cohoes mastodon. So also when, during the San Francisco earthquake of 1906, he calmly measured the burning time for wooden buildings and when, during the hard-pressed Wheeler Survey expedition up the Colorado River, he imperturbably took astronomical sightings on Venus from the gorge. The long search for mathematical forms like the potholes perhaps climaxed in 1908, when he led a visiting geologist into the sequoia groves of the Sierras. Oblivious to the trees, symbols though they were of the wonder and transcendence of the organic world,

Gilbert labored for hours looking for a spider which made its web in the shape of a paraboloid. The greatest wonder of all was that nature was a geometer.

Cracking the Nutshell

By 1868, Gilbert was twenty-five years old. He was employed in a job whose future offered little more than its five-year past. He was not likely to graduate from Cosmos Hall, as did most of its employees, into a curator's job at a museum, and he was not likely to found a similar institution like the "Microcosm" established by his comrade Howell. Cosmos Hall had served its purpose. It was time for the moratorium to end.

During these years the essential Gilbert crystallized into patterns which were repeated throughout his life. In the Nutshell, he had an archetype for all his social relationships. In his investigation of the mastodon, he discovered possibilities in a geologic science quite distinct from that practiced by his contemporaries. He showed a preference for mechanics over fossil artifacts; he sympathized with an inorganic nature whose hieroglyphics were mathematical forms rather than with an organic nature alive with transcendental symbols. Rather than the natural history of Cosmos Hall, he pursued the natural philosophy he had learned at the University of Rochester. Around Rochester, he found a geologic landscape whose themes he returned to time and again. From the falls at Cohoes and Niagara, he acquired a fascination with the erosive power of rivers; from the Great Lakes, whose inlets lapped to Rochester, he was exposed to the processes and shoreline environments of lake bodies; from the Pleistocene topography of upstate New York, he encountered a variety of features whose unity clustered about the climatic changes which produced the glacial epochs.

Gilbert returned to these themes as often as he returned to the Nutshell. This was appropriate. If these themes defined the range of his scientific thought, the Nutshell and his classical education denoted the dimensions of his personality. In the end they made up the horizon of his consciousness, so that the consistency evident in Gilbert was little different from that which he perceived in the world. This meant that in his personality no less than in his science the man who measured the Cohoes potholes stands as a somewhat isolated, even unique figure. From an urban rather than a rural environment, he failed to share that fascination with the organic world (and its petrified relics) that typified most of his associates. He accepted a mechanical metaphor rather than an organic one; he saw equilibrium rather than evolution as nature's pattern; his intellectual mentor was William Rankine rather than Humboldt, Darwin, Hall, or Ward. He came to geological science from a formal education in the classics rather than as a self-taught naturalist. His father was not a minister, as was Powell's, nor was his family famous for its moral fervor. Grove Sheldon Gilbert, on the contrary, had apostatized, and any remaining religious sentiment in his son vanished when GK's only daughter, Betsy, died of diphtheria at age seven. He was more of a scholar than an adventurer, a conservative rather than a reformer; he passed through the Civil War without seeing value to the concept of "struggle for existence" and without becoming a scientific captain of industry. He was an investigator, a mental state he contrasted with that of the teacher. Consequently, he never accepted an academic position or a government podium or wrote a definitive textbook, so that despite his years of intellectual leadership he never established a school of followers or articulated a distinctly Gilbertian creed. He was an unassuming maverick: his reserve rather than his enthusiasm set him apart. The contrast to his major contemporaries is nearly perfect.

It was indeed a strange amalgamation—an improbable mixture of classical languages, the private orthodoxy of the Nutshell, and the scholarly temperament of an unsuccessful teacher. Yet the problem of selecting a career, an imperative which grew more demanding each year, had discovered a possible solution in the Cohoes potholes. What was born in the Nutshell underwent its rite of passage in 1869, when—"with a lot of cheek," as he put it—Grove Karl Gilbert traveled to Columbus, Ohio, to ask Gov. Rutherford B. Hayes for a job on the Geological Survey of Ohio.

2. "Astride the occidental mule"

I have shown how useful, even indispensable, fossils are to the student of geology, and I am happy to know that their significance and value are coming to be generally known.

—John Strong Newberry

The necessities of the War Department demand geographical surveys to be made . . . that work has been done always by our War Department, since the formation of this Government, and I am satisfied that it will continue to be done, and that the wants will become larger year by year.

—Lt. George M. Wheeler

A Volunteer Assistant

Gilbert's request was more than a bid for a job. So far his education in geology, except for occasional episodes exhuming bones and examining potholes, had been confined to reading textbooks, cataloging mineral specimens, and labeling assorted fossils. Neither the principles stated in the texts nor the particles funneled through Cosmos Hall taught the hard methods of fieldwork. Once committed to a career in geology, the twenty-six-year-old Gilbert wanted an apprenticeship even more than simple employment.

The Ohio Survey was an excellent choice. Gilbert's experience with James Hall at Cohoes brought to mind the influential New York State Geological Survey, and it is likely that he encountered John Newberry during the process of excavating and reassembling the mastodon—all of which pointed to a survey job in Ohio, where Newberry was the newly appointed chief. Besides, its Great Lakes terrain was not much different from that around Rochester. There were negative reasons as well: the available options were few. Museums were a possibility Gilbert rejected; he was unsuited for teaching, even if any academic positions had materialized; federal science, except for the Coast and Geodetic Survey and the agriculture service, was largely confined to the military. Academic surveys, like academic science, were just beginning to develop. Although by 1869 O. C. Marsh was shepherding Yale students to Nebraska via the Union Pacific, and even Illinois Normal was conducting informal tours to the Rockies led by an ambitious Civil War veteran named John Wesley Powell, their function was basically to collect specimens for museum study and display. To Gilbert that must have seemed like an antecedent to Cosmos Hall.

Of the choices open, the Ohio Survey made wonderful sense, except that Gilbert was not a native Ohioan and Hayes refused his petition on those grounds. Apparently GK had enough cheek left to turn his other to Newberry. That personal appeal also failed, for the same political considerations. But Newberry's refusal was not unconditional. Perhaps he remembered an enormous skeleton from Cohoes that had been painstakingly reassembled, or maybe he simply needed some inexpensive talent. For whatever reasons, the two men worked out an agreement: Gilbert would work as a volunteer assistant, with fifty dollars per month for expenses. He had his apprenticeship. The incident was as forward an act as Gilbert had yet dared but because of it he was a member of the Ohio Survey, well on his way to superb training as a field geologist.[1]

The Ohio Survey of 1869 built on the practical and theoretical results of its 1837 predecessor, captained by W. W. Mather. But in many respects, in both style and emphasis, it differed. The early survey reflected the politics of the internal improvements movement. A general reconnaissance of natural history, it accentuated as economically valuable those resources which related to agriculture—climate, soil, and potential transportation routes for canals or railroads. In this the state survey paralleled the inventory conducted by the federal government in its rapidly expanding western territories. Like the colleges which proliferated at the same time, the state surveys aspired both to bring civilization to the frontier and to celebrate local resources. Indeed, just as the same man often served as a canal engineer and geologist, so he also frequently taught at a local academy.

However, separating the two surveys were three events: the Civil War, the *Origin of Species*, and rapidly spreading industrialization. The Ohio Survey of 1869 absorbed their consequences. This meant that as a scientific enterprise, the survey described the geologic record in terms of historical evolution; in utilitarian terms, it was translated into an investigation of those resources and manufactures relevant for the state's industrial growth. As Newberry summarized it, the survey was conceived as a means of "repairing the breaches of war, and moving faster the retarded wheel of progress." Perhaps, too, the Ohio legislature was spurred on by the example of the new Wisconsin Geological Survey, under Thomas Chamberlin, or the California Survey, organized in 1861 under J. D. Whitney, where a young Yale graduate named Clarence King joined up, as Gilbert had, in the capacity of volunteer assistant. In any event, the new wave of state surveys constituted an interim mechanism for training and employing a large body of scientists.[2]

Eventually survey problems were nationalized, as well as the survey system. The states reduced the size of their bureaus, while the federal government, along with the universities, shouldered the burden of scientific research and supervised the investigation and regulation of those resources, such as water, which involved more than local boundaries. But, until academic science and the federal bureaus matured, the state surveys were a significant proving ground for the men, like Gilbert, who later staffed the national institutions.

Yet perhaps the most formative aspect of his new job was the association that sprang up between Gilbert and his brilliant chief, one of those versatile naturalists whose personality and enterprise wrote large chapters in nineteenth-century American science. Like nearly all the leading geologists of the century, Newberry was an exceptional teacher; at the time of Gilbert's introduction, he lectured at Columbia University when winter prevented fieldwork in Ohio. From the strength of his dual appointments as professor and survey director, John Strong Newberry helped make many scientific careers. During his tenure with the Ohio Survey, he made Gilbert's.

His new boss was a type GK repeatedly found attractive, and their relationship ripened quickly into respect and friendship. In later years, when Newberry presided over the geology department at Columbia University, Gilbert often detoured to New York for a visit en route to the Nutshell. The esteem was mutual. Newberry found his volunteer assistant to be both skillful and conscientious; Gilbert found his chief to be an excellent teacher. After the end of each field season—each lasted roughly from July to September—Gilbert assisted his boss in New York between visits to Rochester. He helped straighten out the season's collections—he put his years at Cosmos Hall to good use here—assisted in arranging Newberry's winter lectures at Columbia, drafted reports, and drew illustrations—exploiting his father's training in art. The drawings, in fact, were what originally impressed Newberry. He exclaimed in one progress report

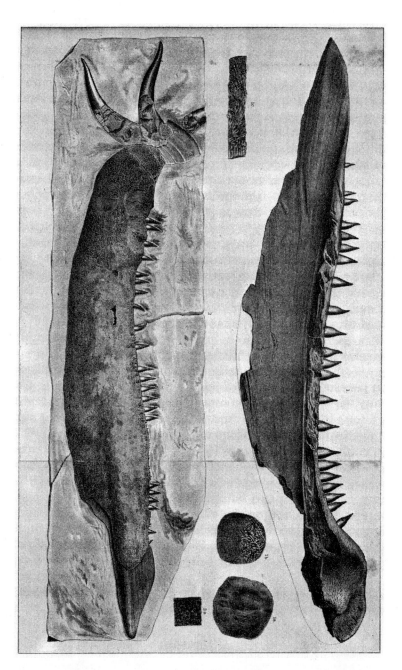

Gilbert the draftsman. It was skilled draftsmanship, as in the mandibles of this Devonian fish, that first drew Newberry's attention to his young assistant. The fish is typical of those which ignited Newberry's scientific and poetic fancy. *Reproduced from the 1873* Report of the Geological Survey of Ohio.

that they exhibited "a style that has not been surpassed in this country, and some of the work is equal to any of a similar nature done by
the best European draughtsmen."[3]

In return for these services, Newberry introduced Gilbert to the
reigning scientific dignitaries of Columbia and Yale, and he broadened his assistant's cultural education and entertainment habits to
include theaters, lectures, and sermons. On one occasion Gilbert
showed his peculiar humor by attending two lectures in a single
day: "One by the eminent Congregationalist, Henry Ward Beecher,
on the 'Request of the disciples for more faith,' and the other, perhaps as an antidote for the first, by that ill-balanced iconoclast,
George Francis Train, on 'Old fogies of the Bible.'"[4] During this period he became active in a local scientific association, the New York
Lyceum of Natural History, and there delivered his first scientific
papers in 1871. Based on material he gathered during his two field
seasons in Ohio, they repeated the themes of his previous articles
from Cohoes: an examination of a river valley and the excavation of
an Ohio mastodon.

No less influential were the associations he made with people
connected with the survey. In many cases their scientific paths and
Gilbert's intersected opportunely in the future. Edward Orton, in
particular, earned Gilbert's admiration. Like Newberry, whom he
succeeded as survey chief, Orton directed the survey and taught simultaneously, serving as president of the newly organized Ohio
State University. In a biographical memoir, Gilbert unabashedly celebrated Orton's talent for "practical science." He performed a different kind of obituary for another member of the survey, Henry
Newton. When Newton tragically died in 1878 while completing a
draft of his monograph on the Black Hills, Gilbert edited the penciled manuscript for publication. Somewhat less dramatically, he remained acquainted with the other geologists on the survey—R. D.
Irving, E. B. Andrews, N. H. Winchell—and with the battery of distinguished paleontologists to whom Newberry farmed out many of
his specimens—including F. B. Meek, E. D. Cope, O. C. Marsh,
James Hall, R. P. Whitfield, and J. J. Stevenson.

Yet helpful and gratifying as these social and professional ties
might be, they amounted to a fringe benefit to Gilbert's two years
on the Ohio Survey. His primary ambition had been to learn practical field geology, a transition he did not find exceptionally difficult.
If he was initially inexperienced in the routine of field investigation,
he was at least broadly familiar with the field. Newberry had divided
the state into several large districts and dispatched Gilbert—perhaps at his request—to the counties bordering Lake Erie. Gilbert

easily mastered the routine. This was not a taxing achievement when one considers that, pasted in the front of the officially issued notebooks, was a detailed listing of "Directions for Collecting and Observing." The catalog was probably as much for the benefit of Ohio legislators as for the volunteer amateurs who assisted the survey, but it made explicit the assignments for each district.[5]

The bulk of the instructions concerned surface deposits and their mapping. Soil and vegetative covers were to be noted for the benefit of agriculturists, while moraine records would serve scientific research by testing the increasingly elaborate glacial hypothesis first announced by Agassiz in 1837. Assignments included the study of "economic geology" (that is, mineral deposits) in conjunction with "geologic structure"; the compilation of statistics on mining and manufacturing; and the collection of geologic specimens and Indian relics—as though the two were equivalent—as much to display as curiosities in public museums as to unsnarl the stratigraphic record. In short, the Ohio Survey was conceived as a scientific inventory of the natural resources and economic potential of the state. Newberry's error—according to the legislature—was to organize his interpretation of that wealth as a portfolio of arcane fossils rather than as a portrait of future factories.

In practice the instructions for observers meant that the survey simply employed geology as it was understood at the time. Some of the directions produced little more than genuflections. Some, such as those dealing with soils, were pursued casually. And, in accordance with orders, field parties dutifully visited a drab succession of factories. Mostly, however, the survey mapped surface features, constructed stratigraphic profiles, gathered fossils, and elaborated the correlations of the regional stratigraphic column, while the legislature waited interminably long for the reports to materialize—only to subsidize, to their horror, a succession of costly tomes on the strange inhabitants of the Paleozoic.

If the survey sometimes failed to accelerate the process of the industrial revolution in Ohio, it benefited greatly from it, particularly in transportation. Survey travels were largely done by railroad car or by walking along the banks of abandoned canals. The rails freighted geologists out into the field and hauled rock samples back. But the tracks of the industrial revolution also served an intellectual function. The engineering reports for railroad and canal construction listed elevations that were useful in constructing topographic maps, and in a region heavily mantled with glacial debris—and that shrouded with vegetation—rail cuts and canals provided a convenient slice of bedrock. James Hall and Amos Eaton had earlier demonstrated how valuable such exposures were for paleontologic and stratigraphic research.

What was true in Ohio proved equally compelling in the Far West. With river gorges taking the place of canals, with a proliferating network of rails to whisk scientists to auspicious locales, and with vast mountains largely denuded of vegetation, it is little wonder that great insights were possible west of the Rockies. It only remained to graft the investigating techniques practiced along eastern canals and railroads onto the waning style of the army reconnaissance to fully outfit the late nineteenth-century western scientist-explorer, equally at ease on a mule or in a Union Pacific diner. It should come, then, as little surprise to learn that one of Gilbert's finest theoretical concepts—the graded stream—took its name from the engineering term used to describe the longitudinal profile of railroad tracks. Moreover, it was in solving the questions raised by the new machinery of steam power that thermodynamics developed as a science and, at least in Gilbert's case, it was by conceiving of the landscape as a similar sort of engine that he made his most original contributions.

Although Gilbert gratefully served as a sort of orderly to Newberry and respectfully applied himself to the stratigraphic riddles that preoccupied his chief, his field notes betray other interests. It is possible that these reflect specific assignments ordered by Newberry, but more likely they reveal Gilbert's own curiosity. The subjects: shorelines in the form of moraine ridges and oscillating water levels of the Great Lakes. His solutions did nothing to economically advance the industrial revolution in Ohio, but they initiated a slow process by which, in other places and at later dates, the American landscape and industrial machinery were joined by striking intellectual analogies.

"Labels written by the Creator": John Strong Newberry

Gilbert's vision of geology was not that of his boss. But it was GK's great good fortune to serve under John Strong Newberry, who, along with Dana, did more than anyone else to weld the Humboldtean explorer to the emerging evolutionary paradigm of the earth sciences. The son of an Ohio coal mine owner, Newberry took a commonly traveled path from study as a physician to a professorship in natural history. En route he studied geology in France, where the science was still awed by the lingering memory of Cuvier. In 1855, his combined skills as physician and naturalist landed him an appointment with the Williamson-Abbott Expedition, a transit run by the

Army Corps of Topographical Engineers between San Francisco and Portland as part of the Pacific Railroad Surveys. When the expedition ended a year later, Newberry set up shop in Washington, establishing ties with the Smithsonian and with George Washington University. Thus was set the fundamental pattern of his career, a happy migration between fieldwork and lecturing.[6]

The Pacific Railroad Surveys were an attempt to solve the political problems of sectionalism through exploratory science. Before long more problems with sectionalism—this time in Utah—sent the army engineers, and Newberry, back into the field. In 1857, he joined the expedition under Lt. Joseph Christmas Ives in the capacity of physician and naturalist. Here was a stroke of luck. The Ives Expedition planned to explore the possibility of using the Colorado River as a supply route for federal troops who, as an assertion of federal sovereignty over a rebellious territory, were already marching across the plains to Utah. Thus the proposed route took the expedition to the western gorges of the Grand Canyon. Like many other army explorations, the Ives Expedition had German scientists and adventurers on its staff. One was F. W. von Egloffstein, a celebrated cartographer; the other, Heinrich Möllhausen, having married the daughter of Humboldt's valet, came with the aged explorer's personal blessing, and Möllhausen's account of his previous adventures with the Whipple Survey had been published with an introduction by Humboldt himself.[7]

The survey's steamboat, *The Explorer*, ran aground in Black Canyon, but the troops still managed to enter the Grand Canyon at Diamond Creek. They completed the expedition by traversing northern Arizona to Fort Defiance, New Mexico, in 1858. The following year, Newberry joined a reconnaissance conducted by Capt. John Macomb from Santa Fe to the junction of the Green and the Grand rivers. When he finished, Newberry had seen more of the Colorado Plateau than had any other scientist and had few rivals as exploration geologist in the Humboldtean vogue.

The reports of the two expeditions brought world fame. The Ives report was published in 1861; the Macomb report, much delayed by the Civil War, appeared in 1876. Newberry organized his material, perhaps of necessity, in the form of an itinerary; yet he successfully drew many important stratigraphic columns (including one for the Grand Canyon), gathered abundant specimens, and summarized in a few pithy remarks the basic facts of Colorado Plateau land sculpture. Its bizarre forms he decided were caused by *"a vast system of erosion, and are wholly due to the action of water."* That remark refired national controversies in European science, where the potential work of streams was considered trivial when compared to marine erosion. Only his fellow American James Dwight Dana had credited streams with a similar erosive power. Indeed, as a result of his years with the Wilkes Expedition, an oceanic version of the continental explorations by the army, Dana became Newberry's only serious competitor as an explorer, teacher, and philosopher of American geology. Together the mineralogist and the paleontologist founded what, by the end of the century, was clearly recognized as a distinctly American school of geology, one of whose central tenets was the power of fluvial erosion.[8]

Its other central tenet was the universal "law of progress" apparent in natural history. It was through this insight that both Dana and Newberry successfully transcended the conceptual limitations of the Humboldtean reconnaissance. Deeply religious, both men at first resisted evolutionary notions, then embraced them. Both Dana, through an analogy to embryological development, and Newberry, through the fossil record, saw in the regular succession of earth events the evidence of historical design in nature. Hence, natural history retained its link with natural theology, a point lost on neither man. Joseph Le Conte in 1895 memorialized Dana's contribution to geology by observing that, "so long as the Lyellian idea of geology prevailed," geology was "a mere field for the application of physics, chemistry, and biology. Geology became one of the great departments of abstract science with its own characteristic idea and its own distinctive method under Dana." As Dana's son-in-law, Le Conte may be pardoned a little exaggeration, but he should have credited Newberry with as substantial a role. It was Newberry as much as anyone else who pointed out the terms of reconciliation in America between evolutionary thought and natural theology. The reason was the design apparent in the geologic record by which "the different ages, periods, and epochs follow each other everywhere in regular order, and form a grand and uniform system of change and progress." It was as self-evident in the natural as in the social world that "this is an age of *progress.*"[9]

What confirmed this belief, and what established the basic equations of geology, was the fossil. Through fossils, rocks could be correlated both geographically and historically. No longer mere sports of nature or a brand of natural obscurantism, fossils had become successively a fundamental enigma, a tool, and finally the language of creation. Moreover, fossil bones, encrusted with calcite or dripping with tar, were a scientific equivalent to the ivy-covered ruins so attractive to the romantic imagination. The explication of fossils was a cultural requirement equal in significance to the explanation of toppled monuments in Yucatán, the pyramids of Egypt, and the sandswept pillars of Ozymandias. The moments of present and past had to be arranged and unified; the medium was history, the mechanism evolution.

As a result, paleontology became the mathematics of geology, and stratigraphy, the first specifically geologic discipline to be organized by fossils, became its mechanics. Biological and geological thought were joined as symbiotically as mathematics and physics. This union, in turn, created an unexpected imperative. Since, to American minds, biological adaptation depended on the character of the physical environment, earth history had to evolve in tandem with organic history, the record of rocks had to parallel the historical tree of life preserved in fossils. Newberry's Ohio Survey, assembled ten years after the publication of the *Origin of Species*, was fully cognizant of that fact.

Hence, when the survey issued its two progress reports and published its completed research in four volumes, the informing principle was evolution. Of the first three volumes, half was devoted to paleontology; the whole of the fourth pertained to botany and zoology. The central geologic theme was stratigraphical—working out a fuller geologic column and correlating Ohio strata to those in the neighboring states. Economic geology (that is, coal beds) appeared as a unit in the column. Visually, the reports were synthesized by geologic maps, stratigraphic columns, and cross sections. Verbally, the unifying theme to the geologic record was—as it had been for Newberry's western explorations—a narrative: the historical development of strata and fossils.

This is not to say that Newberry was disdainful of "economic" deposits. On the contrary, as befit the son of a coal mine owner, he called coal "the mainspring of our civilization" and argued that "its possession is the highest material boon that can be craved by a community or nation." Yet what fascinated him most about coal was its poetry. "It has been formed under the stimulus of the sunshine of long past ages," he wrote excitedly,

> and the light and power it holds are nothing else than such sunshine stored in this black casket, to wait the coming and serve the purposes of man. In this process of formation it composed the tissues of those strange trees that lifted their scaled trunks and waved their feathery foliage over the marshy

shores of the carboniferous continent, where not only no man was, but gigantic salamanders and mail-clad fishes were the monarchs of the animated world.[10]

A romantic vision. But Newberry believed with some justification that those gigantic salamanders, mail-clad fishes, and feathery-scaled trees were what made sense out of the coal, that they explained most of what geology could say about the meaning of coal. It was a position not far removed from that taken by J. D. Whitney in California when he pivoted his remarks on the auriferous gravels upon the discovery of a cephalopod.

Newberry had won a bitter contest to become director of the Ohio Survey by insisting on exactly this point. When his chief competitor, Col. Charles Whittlesey, turned a private feud into a public debate, Newberry, never one to avoid a good fight, retaliated with knuckles bared. Whittlesey was too old, he claimed, too frail, and too ignorant of geology. In particular, he was no paleontologist; and, without a comprehensive knowledge of fossils, geology was gibberish. He made the same point against the "practical men" in the state legislature who didn't "care a row of pins" for his "clams and salamanders." Newberry endeavored to "prove to them that the fossils which they despise *are* eminently practical; that they are labels written by the Creator on all the fossiliferous rocks, and that no one can be a Geologist who has not learned their language." Fossils were the alphabet of natural history; evolution supplied a syntax and made them the language of nature.[11]

On this issue, Newberry won the battle against Whittlesey but five years later lost the war with the state legislature: he was forced to resign in 1874. The discovery of a mastodon, a Devonian armored fish, or an extinct fern could not quite compensate for the failure to locate coal, petroleum, or gold. Like Whitney, among others, Newberry had strategically erred by postponing the economic reports in favor of thick volumes on paleontology. Conversely, Newberry's successor, Edward Orton, steered the Ohio Survey through four more years by writing some masterful treatises on petroleum and natural gas.

The lessons of all this were not lost on Newberry's volunteer assistant, and twenty years later he would witness the cycle repeated on a national scale. The professor-explorer Newberry had cashed in his Civil War experiences with the Sanitary Commission for practical administrative skills, just as Chamberlin and Powell would later do, to become a scientific captain of industry. When his connections with the Ohio Survey were severed, he transferred his organizational abilities to the Columbia School of Mines, where his dynamic temperament, moral sincerity, religious conviction, and scientific competence lent him enormous prestige. From his academic post, he continued to enter frays in the politics of science. His 1879 letter to Rutherford B. Hayes, for example, was persuasive in establishing the U.S. Geological Survey and in deciding among the slate of candidates for its directorship. When he died in 1892, at the age of seventy, his acclaim was international and, for his fundamental contributions to the institutional and intellectual structure of the science, he shared patriarchal stature with James Hall and James Dwight Dana as one of the formative personalities in the American geological tradition.[12]

His influence on Gilbert was no less crucial. He readied him for his glory years on the Colorado Plateau—training him in the techniques of exploration; outfitting him with an array of conceptual tools, like fluvial erosion, rhythmic deposition, and local crustal warpings, that reappeared in Gilbert's writings; and securing a place for him on the Wheeler Survey of 1871. Gilbert's two years of apprenticeship with Newberry in Ohio foreshadow closely his more significant relationship with John Wesley Powell. Another professor-explorer, Powell piloted a national survey and lectured to a national legislature, but he also found in Gilbert a valuable adjutant. Powell's explorations, as did Newberry's before him, threw out intuitions that others, notably Gilbert, nurtured into exhaustive monographs. Unlike their studious assistant, both men were really all-purpose naturalists, visionaries of scientific progress, and their exploration narratives, like rock strata stocked with precious fossils, amounted to adventure stories studded with gems of scientific insight.

Their remarkable similarities, however, also accentuate the fundamental intellectual distance separating them from Gilbert. Basically, he disagreed with the role of the fossil. Already trained in mathematics and mechanics, he felt little need to work with their natural history equivalents. He knew the taxonomic languages, but the references to fossils in his field notes, as in his reports, were rare. Time and again, he reaffirmed his preference for Lyell's steady-state world over the universal laws of progress that Newberry and Powell found written in large script. He would have disputed with the embittered Colonel Whittlesey, who, on the basis of his experiences with Newberry and Hall, had published the suggestion that "the study of fossil remains" had "a demoralizing effect." Yet on one page of his notebooks is a drawing, almost Kafkaesque in its conception, which may best characterize his attitude toward the debate. It is the figure of a man with a trilobite for a head.[13]

"Hydrographical peculiarities"

That caricature speaks volumes about GK's geologic interests in Ohio. His official reports on the geology of individual counties dryly followed a standard format. Each of his "Directions for Collecting and Observing" merited a separate discussion, which usually meant a catalog of tree species, Indian artifacts, building stones, strata, and so on. In the course of this automated reconnaissance, however, he became intrigued by the surface sculpture of the Maumee Valley. In fact, its "hydrographical peculiarities" as viewed from a map were "so singular and striking, as to have excited some attention and curiosity before the region was visited." In a special chapter inserted into the Ohio Survey reports, an address before the New York Lyceum, and an article in the *American Journal of Science*, Gilbert worked out an interpretation for that eccentric topography. In one sense, his glacial chronology supplied only the final crust to the historical evolution of the landscape. In another sense, his analysis was no more a simple exercise in glacial stratigraphy than his study of potholes had been an essay in paleontology.[14]

In the Maumee Valley, Gilbert examined ridges, which he interpreted as terminal moraines; terraces, which he showed to be old shorelines of Lake Erie; and wave action, for which, with help from E. B. Andrews, he formulated a general law—"wherever the shore current is accelerated the waves cut away whatever opposes them, and wherever it is retarded they accumulate sand." He searched for outlets draining the ancient lake. He described a theory of glacial action in which grooves resulted not from gouging boulders but from "the tendency . . . of glacial ice to prolong a resisting knoll into a ridge and a cavity into a groove"—a product of the peculiar stresses of ice flow acting on rocks of unequal resistance.[15]

The Maumee Valley streams formed broad bows, roughly concentric to the present shoreline of Erie. These were integrated by a major channel draining into the lake. After careful examination, Gilbert discovered a series of small terraces, which he identified as terminal moraines reworked by an older, larger lake. These terraces formed a barrier which, with the retreat of the lake, could not be breached by the new streams which attempted to flow into it. Only where meltwater from the ancient glacier had reached the moraine could the new runoff drain into the lake, and precisely along this axis was the main channel of the Maumee River. Gilbert showed that, when the glacier first began its retreat, the lake impounded between it and the moraine rim had drained southwest through this channel. Yet he also recognized other former outlets and, borrowing a suggestion first published by Newberry in 1862, he concluded that there had been some local crustal motion following the glacial retreat. "There seems to me to remain but one hypothesis, and that the true one," he wrote:

> that the warping of the rim of the lake basin has taken place; that there has been local subsidence, or upheaval, or both, so that Fort Wayne is now relatively much higher than formerly; and that it has been by the contortion of the great reservoir, that the point of its outflow has been shifted from point to point, and its capacity has been varied.[16]

While the prose may have been carpentered a little self-consciously and the first-person voice may have quavered somewhat, Gilbert's logic was as irresistible as the glacial advances he described. But what made this more than a lucky hunch were the calculations in the text. They showed, for example, the differences in elevation between the several outlets. Gilbert vividly sketched the consequences. "To restore now the old water level and current at Fort Wayne," he noted,

> we would need, not merely to fill the gorge of the Niagara, and renew the escarpment at Lewiston, but to construct on that escarpment a retaining wall 170 feet high and many miles in length; and, after filling the valley of the Desplaines to the height—one hundred feet—of the adjacent drift hills, another hundred feet would be needed to complete the dam.[17]

In its way it was a brilliant image, but one philosophically as well as stylistically far removed from Newberry's romantic justification of the poetry of coal. Already Gilbert's early attraction to engineering showed itself in his use of concrete visual imagery, like retaining walls framing the old lake.

Not everything GK said proved correct. Newberry gently commented in a footnote that some beach terraces probably resulted from the lake waters created by the ice dam made by the retreating glacier, not simply from crustal warpings which dammed outlets, and Gilbert adjusted his later work to incorporate that shrewd insight. Interestingly enough, even at this date he accepted Lyell's long-influential iceberg mechanism for transporting "glacial drift."

What makes this fact doubly curious is not only that Gilbert clung longer than he should have to Lyell but that he gave no evidence of having read Darwin or Dana. This paradox holds throughout his career. At one point in his text on the Maumee Valley, he gave an indirect explanation for it. "We may read once more in this the lesson, which modern science teaches in so many ways," he wrote, "that the present is but the continuation of the past; that geology, as well as the history, is now enacting; and that it is only because of the brevity of the time allowed us for comparison that nature seems to have reached or approximated an equilibrium."[18]

Yet it was precisely that "enacting" present, that apparent equilibrium, which attracted Gilbert and made him cite Lyell, whom he admired and read closely, rather than the evolutionary geologists. With his documentation of the processes working on the earth at the present time, with the conservation principle which underwrites, for example, his depiction of "crustal undulations," and with his reluctance to admit developmental change through time, Lyell strongly appealed to Gilbert. Hence, the volunteer assistant could afford to draw cartoons of trilobites. His field of interest was the landscape of the present, not of the Paleozoic. In the conclusion to his analysis of shifting outflow channels in the Erie basin, Gilbert acknowledged the complexity of the question involved. But he announced—and in the Great Basin twenty years later he proved—that "the problem is not insoluble."

At the end of his essay on the geology of the Maumee Valley, Gilbert described a fossil mastodon excavated near the village of Saint John. He analyzed closely the circumstances of its deposition, fixed in a bog. In fact, the riddle of its mechanisms of deposition proved more interesting than the question of its age. In the end, rather than use the fossil to date the stratum, he used the physical conditions of the deposit to date the fossil. That, too, was a lesson he transported to Lake Bonneville, and it was also a reason why, for all his personal attachment to Newberry, he never published a study in historical geology or a beautifully illustrated thesaurus of Devonian fossils. In its way that insight was a twist as significant as those bends in the Maumee Valley streams that first aroused Gilbert's curiosity.

At the end of his two-year hitch, it was less Gilbert's peculiar curiosities than the totality of his work in Ohio that really mattered. He impressed his chief well enough that, when the army geared up in 1871 for a second assault against the Colorado River, Newberry recommended Gilbert for the position of geologist. Just as

Lt. George Wheeler, the survey's commander, stood as the hopeful successor to Ives, so Gilbert now stood in relation to Newberry: in this second wave of federally sponsored western exploration, Gilbert took a role similar to that which Newberry had assumed in the first. In recapitulating the formative experiences of Newberry's career, he was introduced to the terrain and themes he would make especially his own, to a landscape whose "hydrographical peculiarities" were magnified on a grand scale.

GK occasionally regretted that he had begun his career rather late in life. If one takes his work at Cohoes literally, he had indeed begun in a hole. But now, at age twenty-eight, his apprenticeship was behind him. Outfitted with the basic techniques of field geology, with copies of Newberry's old maps, reports, and brilliant hunches, with a substantial education and a special vision of natural phenomena, Gilbert was en route to a spectacular career as an exploration geologist in lands far distant from the Nutshell. On May 7, 1871, amid snorting mules, blowing alkali, and blaspheming sergeants, he arrived at Halleck Station, Nevada, to join the U.S. Army's Geographical Survey West of the 100th Meridian.

West with Wheeler

The force which assembled at Halleck Station was a large one: over forty men—the bulk of them military escorts—and 165 horses and mules. The group took shape as a military caravan. Its size, however, was only commensurate with its ambitions, for the Wheeler Survey (as it came to be called) was attempting to recapture the momentum which the army had lost to other exploring groups by reinvigorating the antebellum methods that had formerly proved so productive and so glamorous.

Yet its competition was formidable. By the time Wheeler assembled his challenge, there were three other federal surveys mapping the Far West. One belonged to Clarence King, who in 1866 conceived the idea of exploring a hundred-mile swath from the Sierras to the Rockies, a route he cleverly drew through the tracks of the Central Pacific. Though nominally under the direction of Gen. Andrew A. Humphreys in the Army Corps of Engineers, this was strictly King's show, and at age twenty-five he conducted an impressive band of well-educated geologists and mining engineers in the Exploration of the 40th Parallel. Meanwhile, approaching the Rockies from the plains, under the auspices of the Interior Department, was the Geographical Survey of the Territories. By 1871 the Hayden Survey, as it was called, was an omnibus but ambitious affair. By contrast, the Powell Survey, which sprang out of his stunning 1869 descent of the Colorado River of the West, was almost a family affair. The two daring voyages had made Powell a national celebrity, however, and by 1871 he sought to bring his survey into similar prominence. Operating under the grandiloquent title of United States Geological and Geographical Survey of the Rocky Mountain Region, his crews fanned out from Kanab, Utah, for exploration and mapping. Collectively, this civilian onslaught threatened to overshadow its military rivals.[19]

But the army was not in total eclipse. It had continued reconnaissances as they were needed, usually to gather intelligence about territory occupied by hostile Indians. Yet the exploring army after the Civil War resembled the fighting army: it seemed banished to trivial rounds of routine duty, facing a Congress as unfriendly as the Indian tribes it endeavored to suppress. Still, its exploring tradition was a proud one, and spearheaded by Humphreys the army fielded a small battalion to compete with its civilian challengers. With two years of reconnaissance leadership behind him, tours of Nevada in 1869 and 1870, Lt. George Montague Wheeler was chosen as the army champion.

Not surprisingly, the 1871 expedition that Gilbert met in Nevada proceeded as though it were on military maneuvers. Under the guidance of General Humphreys, Wheeler had carefully plotted his routes in advance, periodically issuing and updating the itinerary in the form of orders. He broke the company into smaller parties, usually two, which were, in turn, commonly subdivided, and established dates and places for rendezvous. It was a precise exercise—far too ambitious and exact, it turned out—but its purposes were specific and its instructions straightforward. In the 1871 orders of General Humphreys, "The main object of this exploration will be to obtain correct topographical knowledge of the country traversed by your parties, and to prepare accurate maps of that section." At the same time, the Wheeler parties were expected to "ascertain" practically every phenomenon of potential military value and to make general collections in natural history. Mapping, nonetheless, remained the primary object of the survey for its duration. Maps were something the army needed; their construction was something army officers were trained to do; they revived memories of old army pathfinders; and map making offered an alternative to a brutalizing tour of frontier garrisons.[20]

Gilbert was the only geologist on the expedition but not the only scientist or civilian. Archibald Marvine, a brilliant graduate of Harvard's Hooper Mining School, was hired as an astronomical observer, though he and Gilbert frequently interchanged roles. H. C.

Yarrow participated as a physician-naturalist and eventually supervised all the strictly scientific publications of the survey. Oscar Loew collected mineral specimens. Other specialists were added in subsequent years, but in purpose and composition the survey had more in common with the Ives Expedition than with Wheeler's civilian rivals. His civilian scientists were ancillary, their function to collect and catalog, existing parenthetically for "associated natural history observations." Wheeler did emulate his rivals by respecting the power of publicity: he hired Frederick Loring to send colorful copy to the newspapers and magazines.[21]

"Drawing of mules," wrote Gilbert. "Burnt fingers and a dead faint. Dose of whiskey and quick revival."[22] He had a lot to learn. Tall, somewhat thin-chested, fully bearded, reserved—he must have looked a little forlorn and distracted among the frenzy of men and mules. Already the expedition had fallen behind schedule, as officers wrestled with stock and guides, packers with rations and instruments. It was an abrupt initiation into western fieldwork. But Gilbert was imperturbable; he knew his preparation under Newberry had been sound, and his clear-eyed serenity was already as distinctive a trait of his personality as his beard was to his appearance. Yet from the very beginning he knew that this was a magnificent adventure as well as an extraordinary scientific opportunity, and his field notes show subdued excitement. Slowly the unwieldy expedition—it was always too large, Wheeler admitted—moved south.

Late May, the weather cold. "We woke to find the ground covered by 2 inches of wet snow and snow still falling," GK jotted into his notebook, "but managed to trace the road and marched 30 m. . . ." Gradually the unfamiliar became more routine. While the main parties mapped, Gilbert studied creek beds, drew panoramas, and made structural cross sections through the strange mountains of the bleak Nevada desert—"like an ocean wave just about to break and curl westward." There were astronomical observations to make and specimens to label and ship. Yet surprises were abundant: "Killed a rattlesnake, 2 ft. and 3 in., 6 rattles—my first."[23]

The regimen was demanding. Gilbert usually controlled his own daily routine, but the pattern of rendezvous and the pace of the topographic parties dictated his larger schedule; thus he often had little time for detailed inquiry. Fortunately, the barren cliffs allowed quick, visual summaries of stratigraphic relationships and forced examination instead into broad structural considerations. But geology—even as Humboldtean natural history—was not the objective of the survey; nor was its goal simply to map. The army needed a spectacle to match King's Sierra adventures or Powell's river dramas. In his inordinately long itinerary for 1871, Wheeler had two sites he hoped would yield that sort of publicity: Death Valley and the Grand Canyon.

> *Sunday, July 30, 1871.* Memorable on account of a march in Death Valley. Our camp of yesternight was half-way down a winding canyon in the Grapevine range. We had been directed overnight by Kohler [their guide] to Grapevine Springs where we were to connect with the main party. Descending the canyon we visited a highly alkaline springs at the bottom of its wash and had a conversation with three French prospectors whom we thought to be tribes of Wheeler. Observed barometer at sunrise. We were near sea level.[24]

Wheeler's was hardly the first party to enter Death Valley, so he had little claim to priority. The midsummer heat, moreover, precluded much scientific work, while their journey boiled down to a contest of sheer endurance. In Gilbert's words, "anxiety of search on account of water spoiled the day for geology. In the early AM there was no wind or not much about 9, a breeze sprang up that gradually increased in violence until 2 PM the air was full of alkali dust (which we luckily escaped)."[25]

There were some ugly stories about the trip. A few local newspapers wrote sensational copy, but it was not the sort of publicity Wheeler craved. His guides had deserted, maybe had been murdered; he had ruthlessly tortured Indians into selling mules, and so on . . . Gilbert says nothing about this. Yet, true or not, Wheeler's was a personality that could collect this sort of rumor. And the trip *had* been rugged. Even Wheeler dramatized this point when he wrote that

> two of the command succumbed near nightfall, rendering it necessary to pack one man on the back of a mule to the first divide on the route, where a grass sward was reached at the end of a long sandy stretch, while the second, an old and tried mountaineer, became unconscious for more than an hour in nearly the same locality.

Loring nearly joined that fraternity. "The thermometer stood at one hundred and twenty degrees," he wrote for the readers of *Appleton's Journal*, "and I just dismounted from my mule in time to avoid an attack of sunstroke, such as prostrated three of my companions." Gilbert expressed the same sentiment in his own modest way.

Camped for the evening on the valley floor, he wrote nothing but a solitary, understated entry on one page of his notebook—one single, ornately decorated word placed squarely in the center of the page: "Breeze."[26]

The next day "after a dismal march of 3 or 4 hours we found the trail of the mule train . . . and soon water and breakfast. Readied camp at 11:30 and found a party starting to leave at once for Independence direct." Wheeler meanwhile had marched his party for twenty-three hours, some of it "by moonlight, without trail or guide," until they reached a stream gushing out of the Inyo Mountains. The following day, Gilbert's party "marched from Grapevine Springs northwest to the head of Death Valley (proper) and made camp in a mountain belonging to the western (Panamint) range . . . Mr. Shaw from Gold Mountain has acted as our guide today and did not underestimate the distance." The remaining march to Camp Mohave was only a little less exhausting. An evening shower and a spring oasis cooled them temporarily, but the heat blasting off the bare rock cliffs assaulted them. At one point, Gilbert's mule "gave out with hunger and fatigue"; at another: ". . . toward night Lyons came in . . . and reported that he had found hell. 4 miles from Salt Spring senseless and lying on his face with his mouth half full of sand. It was ½ hour before we could get him to swallow and ½ hour before he could speak."[27]

It was not solely an exercise in stamina. Gilbert was too instinctive an observer not to comment on the character of the country, yet most of his ideas came "chiefly from the washes we traverse." Even the sunlight he turned to a scientific purpose: on August 19 he made his "maiden attempt at solar calculations." The year's adventure, however, was just beginning. Thus, on September 15, all parties assembled at Camp Mohave on the Colorado River. The next day, "at 11:35 AM after having posed for a stereograph," the Wheeler expedition began its ascent to the western gorge of the Grand Canyon.[28]

The expedition divided again—this time into land parties and boat companies, all to rendezvous at Diamond Creek. There were three boats with eight to nine men each, plus one barge holding a party of eight. Gilbert traveled with the group which included Timothy O'Sullivan, the expedition's photographer. O'Sullivan had served his apprenticeship under Mathew Brady and broke in as an exploration photographer on King's survey; in turn, during their trials through the gorge, he trained Gilbert. In fact, Gilbert did much of his geologizing in the form of photography. As he editorialized in his final report, "to photograph the complicated is as easy as the simple; the novel, as the familiar." But even photographic geology was often trying. When the wind dropped, the men were forced to row or to drag the boats by towline. When the wind rose, "it interfered with photography and kept O'Sullivan in a perpetual state of profanity." Somewhat the same was true for traditional geology. "Navigation occupied so much attention," Gilbert lamented, "that I had little for the rock." In short, by entering the Colorado canyons with the same methods it used to traverse Death Valley, the survey met with similar consequences.[29]

The boats leapfrogged north. Gilbert was "a little disappointed in the Black Canyon as I had based my ideas on Ives view of the entrance of which I cannot find the original." This was not the last incongruity he discovered between the shape of the land and the gothic drawings of the Ives report. One afternoon, some mountain sheep caused a minor uproar, while soldiers, scientists, and Indians leaped for their guns, and the sheep jumped imperiously to safety. Another escapade soon followed:

> While supper was cooking a mysterious object appeared floating down the stream. Opinion was divided as to whether it was an Indian or a box and it was hailed and shot at in the division of sentiment. Finally after a brisk excitement of ten minutes it was . . . found to be an empty barrel of bacon discarded by the Main Party above us. This ended the second *Battle of the Kegs.*

Then it was the Indians' turn:

> Mohaves and Piutes last night talked, laughed and gambled with each other in our camp. Today after we were well under way Eelitah announced that he had overheard the Piutes planning the shooting of the Mohaves from ambush up the river. So we had a mild panic . . . and concluded to make camp with the other boats tonight though it cost us two hours of fair wind to wait for them to overtake us.[30]

Nevertheless, by October 4, the river parties rendezvoused at Grand Wash Cliffs, which, Gilbert remarked, had "the right to all the adjectives (except numerical) that have been given to it." The next day the land parties joined them, and GK enjoyed mail and a long talk with Marvine. The expedition rested briefly, then pressed on according to its taxing itinerary. Redividing his troops, Wheeler sent half overland to Truxton Springs, where he knew from the Ives

The scholar as explorer in the Grand Canyon. Gilbert geologizing while others sleep, 1871. Photograph by Timothy O'Sullivan. *Courtesy of the National Archives.*

report that Diamond Creek could be entered. The rest he distributed among three boats, carrying rations for fifteen days, with himself, O'Sullivan, and Gilbert in command. Gilbert's years on the Genesee River were paying off. O'Sullivan named his craft *The Picture;* Gilbert in a fit of whimsy named his *The Trilobite.* The system of orders and rendezvous continued to operate, so the boats ran independently for stretches of river, though, especially as the journey lengthened, they were never far apart.[31]

Gilbert geologized whenever possible, trying to work in the early morning hours before the boat parties stirred themselves for breakfast and, during the day, trying to gain time by taking the lead boat whenever he could. One ascent near Grand Wash—"the hardest climb I ever undertook," he wrote—brought him up to a shelf after dark. That night he

> spent along the mountain at the foot of Big Canyon. Having no blankets I built a little fire in a sheltered spot among the rocks and hugged it all night, getting little sleep. The first thing that daylight showed me is that I am on only the first terrace . . . Reached camp about 1 PM . . . Whiskey coffee and rest brought me around however.

He had Newberry's maps and stratigraphic columns to work from. He revised where he could, coining the apt term "Redwall" to describe the domineering limestone cliffs in the gorge, and collected "rounded pebbles from potholes." Less decorously, he termed another stratum "old Snuffy." For the most part, he found the geology monotonous; most of his energies were invested in the strenuous ascent of the river, the residue of his time in his instruments—the camera and astronomical equipment.[32]

At last there occurred one of those baffling disasters that seemed to hound the expedition. The lead boat swamped, then upset, spraying its cargo into the river. Gilbert started at once with a handpicked crew to salvage the floating debris. "We managed to pick up blankets and baggage along shore," he wrote in his notebook with a tired scrawl that night, "but 3 or 4 beds, 3 prs. saddlebags and a considerable amt.—nearly all—the rations were lost." Then followed a calamitous discovery: "The most serious losses were those of the record of astronomical observations of party N 1. and of Lt. Wheeler's basket of papers and notebooks." The major scientific justification for the survey tumbled down a river "the color of red clay and thicker than the Missouri." Gilbert's losses were considerably less severe: "the Macomb Expedition map that Dr. Newberry gave me."

They pushed on. Gilbert met the physical challenge splendidly, quickly asserting himself as one of the leading boatmen. The river parties divided: one boat went "down stream with dispatches and exhausted and demoralized men. The Picture and Trilobite go on up with 10 men (7 white and black; 3 Indian) each." There were some lively transits for the boats that pushed bravely upstream. Gilbert described one which "involved a run out with the line on the Trilobite and then a jerk ahead after throwing it off. The boat was often so highly inclined on a fall that to go forward one must climb as though on stairs."[33]

The rapids worsened. The party was "compelled to make a portage of boat as well as freight." At a double rapid the towrope broke and Gilbert drifted downstream, to begin it over the next day. On October 16, nine days after beginning the drive from Grand Wash, Wheeler put the crews on a short allowance of flour. Gilbert surveyed the ration situation: "Our bacon is gone, and beans, and rice are scant; but coffee is in plenty and will outlast every other item. Our flour will hold out at this rate six days and those must bring us to the Diamond River, or back to the crossing—the former if possible."[34]

The work slowed as men and boats took a daily pounding; the boats had to be caulked, the men sent back downriver or left in camp "to make beds and coffee." Gilbert continued his astronomical sightings. Four miles further upstream, the party underwent "another day of the same sort": there were two accidents, and Gilbert was involved in both. While cargo was being loaded into *The Trilobite*, the fastenings gave way, and the boat "fell backward over the rapid, bumping her stern severely over the rocks and starting a rapid leakage. Richardson, Hoagland, and I had the ride down and did not enjoy it." Three carbines were lost in the spill, one of them Gilbert's. "I do not feel very sorry," he wrote in closing the episode. Later that day, "at a very lively rapid," the party tried to drag up a boat which was empty except for Gilbert and another man. The boat swamped and the two men swam to a nearby shore. Hanging his soaking clothes on a rock to dry, Gilbert pondered "the inconvenience of having no change of raiment."[35]

By October 18, Wheeler was becoming desperate. The expedition was manifestly not turning into a glamour trip. He ordered two men to leave early the next morning for the rendezvous at Diamond Creek. In Gilbert's words, "they carry a demand for grub." GK meanwhile had sighted Venus that morning "just over a 1000 ft. cliff that occluded the sun. This goes well with the bat matter in describing gloom of the Canyon." The next day he and Wheeler took the helms for a grim dash to the rendezvous.[36]

Wheeler's gambit was unnecessary, as it turned out: "By pushing we made D.R. by nightfall." The next day everyone rested. The river trip, Gilbert wrote with considerable understatement, "has proved very exhausting and after 24 hours of nothing to do (heavier than solar observations) I still feel as though just out of a threshing machine." The mule train arrived early in the afternoon.[37]

Wheeler clearly regarded his exploits on the Colorado River as the high-water mark in the history of his survey. In his 1889 summary report, he announced somewhat presumptuously that "the exploration of the Colorado River may now be considered complete." Writing its itinerary for publication in 1872—in a style with one eye still on the Ives report—he remarked with pride that the "river trip, occupying only thirty-three days, was quite an exploration of itself." But he was forced to add that "subsequent revelations showed how inadequate was this (the original) plan, and also the chances for suffering that may arise from want of careful judgment and forethought." Their laborious traverse had succeeded only "after many difficulties, in comparison with which any other of the hardships and privations of the expedition sink into insignificance."[38]

Yet miles remained for the weary explorers. The parties reformed, and, following both Ives and a chain of Santa Fe Railroad stakes, they moved southeast toward the San Francisco Peaks, a chain of extinct volcanoes already adorned with the names of earlier army explorers of the region. The party used guides, and most of the features were already named and mapped. The journey was not without compensation, however: Gilbert learned the virtues of bearsteak—"either as nice or nicer than the finest Hoffman House beef steak. Tender. Juicy. Sweet. Cooked in its own fat." Snow flurries were common. The party detoured to Prescott—"the home of the scum of the earth, morally and financially." Swinging north, they scaled the San Francisco Peaks. Then they began a long loop east and south, following a rugged route not much different from that Coronado had taken over three hundred years before. On December 3, after a final march of twenty-six miles, the terminal rendezvous was reached at a corral outside Tucson. Gilbert had come a long way from the Nutshell. After an expedition which, in Wheeler's proud words, had included marches "from fifty to sixty and even eighty hours, with scarcely a single halt," Gilbert had only one comment: "18 miles the first day and vastly pleasanter than to make 80."[39]

The survey had been a Herculean undertaking. Wheeler loftily noted that "a reconnaissance line of 6,237 miles has been traversed or nearly twice the distance from Washington to San Francisco." Yet little original had been accomplished. Death Valley had been crossed more dramatically by the Briers party in 1849, numerous times since then. Powell had twice navigated the Colorado River downstream; Wheeler's arduous ascent had taken him through the least interesting portion of the Grand Canyon and had then brought him only to Diamond Creek, where Ives had descended on foot.

Wheeler unintentionally admitted this in his 1872 report: "Although the day of the pathfinder has sensibly ended in this country, still it is expected that among the results of an exploration there should be something new."[40] That was precisely the expectation he failed to satisfy. If anything, his labors at the Grand Canyon, for all their adventurous conception, only reconfirmed the fact that, ever since Ives' steamboat, the army style of exploration had failed to navigate canyon geography. His tour through northern Arizona added little; neither did his journey from Fort Apache to Tucson. True, some rough military maps of these areas resulted, but by that time the Apache menace lay mostly along the Mexican border. Wheeler's own early reconnaissances in Nevada preempted his 1871 work. As it grew in complexity and refined its techniques, however, his survey realized a number of valuable military functions; yet in the final analysis it served a purpose greater than its 6,237 miles: the army was firmly back in the exploration business.

Gilbert stayed with Wheeler a total of four years: the first three in the field, the fourth in Washington writing reports. The 1872 and 1873 field seasons—in Utah, New Mexico, and Arizona—were organized in rather the same mode as the 1871, although they were less exhausting in their schedules and more productive in their geology. Wheeler had both refined his system and generalized his purposes. It is, for example, at this time that the survey acquired its full and final title. As a mapping expedition, the survey was better controlled; and, in his justification for maps, Wheeler reiterated their fundamental utility: essential for military operations, yet invaluable for industry (that is, mining) and agriculture (that is, irrigation). While his engineers concentrated on establishing astronomical monuments for triangulation, Wheeler directed his geologists to examine the mines and water supplies of Utah closely.

Basically this meant Gilbert. Archibald Marvine had left for a job examining Lake Superior copper deposits with Raphael Pumpelly; the following year he joined F. V. Hayden's survey. In his place, perhaps on Gilbert's suggestion, Wheeler hired E. E. Howell—an old comrade of Gilbert's from Cosmos Hall who stayed for two years. Henry W. Henshaw was also added as a collector in natural history, and so were others in 1873—military physicians who doubled as naturalists. Gilbert was easily the most competent of the staff, and his earlier training was put to good use.

The operation was taxing but largely routine. GK enjoyed enough time to practically turn his field notebooks into illuminated manuscripts, embroidering the rubric announcing the date into novel forms daily—apparently this brought the same delight he constantly found in elaborating simple card games into eccentric forms. The rationale for surveying was to unite the maps from King's survey through northern Utah with those Powell was drawing in the south. Consequently the parties, broken into five topographic units, spread south from Salt Lake City. Gilbert lingered in the north long enough to inspect the major mines and to measure the terrace levels produced by a Pleistocene lake which he named Bonneville. Then he too turned south.

The High Plateaus of southern Utah were frequently traversed, from the Pine Valley Mountains in the west to Bryce Canyon in the east. Often Gilbert stayed at homes of Mormon settlers, though, on the whole, he found their condition so impoverished as to make little difference whether he camped indoors or out: "They treated us very hospitably (for pay) to what was probably the best they had though it was *very* poor and gave us . . . the floor to sleep on. For this the table and chairs had to be turned outside." A few days later he "lodged with the distiller in Toquerville, sleeping with the son of the proprietor in a covered wagon box, on straw, under a saddle blanket, with a dog on guard."[41]

For a long while the party bivouacked at Kanab, where, interestingly enough, Powell was also camped and where he had run his base line. Gilbert joined William Bell, Wheeler's new photographer, on forays through the incredible landscape around them: southeast to the Kaibab Plateau; south down Kanab Creek to its junction with the Colorado River in the Grand Canyon; east to Lees Ferry, where they "reached camp after sunset and dined by invitation with Mr. and Mrs. Lee (John Doyle). The dinner was a surprise—bountiful and well cooked. Mrs. Lee of 'Lonely Dell' is a black haired English woman of beauty and nerve. The bright kerchief she wore on her head gave a gypsy-like appearance and the stories of her adventures with impudent Navajos coming to cross at the ferry are thrilling in the extreme. John D. fought his battles over again." After they left, the mules all threw shoes, and then "Betsy" threw Gilbert. But for the most part Gilbert's notes for October 24 summarized much of his travels: "A weary ride from Workman's Ranch to Pipe Spring, seeing what I had seen before and losing in the darkness what I had not seen."[42]

On November 25, 1872, Gilbert began a march the length of Utah, up the central valley occupied by the Sevier River. On December 10, he rendezvoused with Wheeler at Salt Lake City. Nine days later, he was in Rochester.

The next year was much the same: in the field from May to December. The terrain was different—western New Mexico and eastern Arizona—and so was the rationale for mapping—to supply the military with the maps it needed to campaign against stubborn Indian tribes in wild country. J. J. Stevenson, another graduate of the Ohio Survey, was added to the corps but, since the main expedition was proportioned into three field parties, this meant one geologist to a survey unit—Howell, Stevenson, and Gilbert. Wheeler felt his scientists had "the opportunity to acquire the most comprehensive ideas of the character of the country." For him, productivity seemed to result in direct proportion from the number of miles logged; needless to say, his geologists thought the relationship should be the inverse.[43]

Most of the rendezvous were at military camps, predominantly Fort Apache. Forests clothed the scenery, so the geology was more difficult to decipher quickly. Still, Gilbert mastered the panoramic sketch: from a mountain perch he could resolve volumes of geologic description on structure and stratigraphy to a few synoptic diagrams. Game and fish were gloriously abundant, yet a bear hunt never fully deterred him from his science. Even at Camp Bowie, Arizona, where he had a "rattlesnake for breakfast; the first I have killed this year," he pondered Baron von Richthofen's theory of volcanic extrusions during the meal.[44]

The work was more prosaic in 1873. Nevertheless, there were plenty of adventures, which were more typically misadventures. On a bear chase in the White Mountains of Arizona, Gilbert's party became separated from the main body. As a consequence, "we five with but one coat prepare to keep a fire all night. The soldiers lie on their face and sleep well but I have little sleep. In the morning we are to start for the trail of the packtrain."[45] At each rendezvous at a military camp, they drew rations: flour, salt, coffee, evaporated tomatoes, beans, bacon, ham, pickles, pears, yeast—all supplemented, to be sure, with local game. Thus outfitted, they circumscribed a great loop, traveling from Santa Fe to Fort Apache (passing Mount Taylor in the process); then they continued south to the Chiricahuas, east to Silver City, and finally north across the Mimbres Mountains and the Gila River back to Santa Fe.

In 1874 the Wheeler Survey again expanded, in both scope and personnel. Gilbert remained in Washington to elaborate his spotty progress reports into a general monograph. Much to Wheeler's dismay, however, the work proceeded slowly, even reluctantly. It was fodder he needed badly, for that year he faced, along with the other

At Apache Lake, White Mountains, 1873. Gilbert is standing to the left, and as usual he is looking in a direction opposite to that of everyone else. The others pictured are a hunter and two Apache scouts. *Courtesy of the National Archives.*

western surveys, congressional review in the form of the Townsend Committee. The delay, he felt, had "allowed of a mis-construction as to the character and amount of the geological work accomplished by the Survey." Gilbert's behavior baffled Wheeler, yet it is easy enough to understand. Its meaning is contained in a jotting to his. field notes which never entered his final report. At the start of the 1872 field season, while still in Salt Lake City, he wrote: "Met Major Powell." He had continued to meet Major Powell more and more in the years that followed.[46]

"A systematic approach": The Geology of a Reconnaissance

The Geographical Survey West of the 100th Meridian published its findings in a stream of annual progress reports, a series of eight volumes of monographs on particular themes, and two atlases, one geographic and one geologic. Although the fieldwork was terminated in 1879, Wheeler retained a staff to finish processing his field data until he released a final, summary volume in 1889. Gilbert's contributions proceeded in analogous fashion, with a slim series of progress reports printed between 1871 and 1874 leading to a massive digest published in 1875. Less formally, he delivered a spate of short papers to local scientific societies and attempted, amid official frowns, to place some articles in the *American Journal of Science.*

His 1875 contribution to the Wheeler report was GK's longest piece of sustained scientific writing to date. In its pages one can see the evolution of a style of thought and of prose as it struggled to encompass the wide vistas of geology hurriedly thrown before its author. More than an essay, Gilbert's major report was still less than a monograph; its shape is that of an anthology of critical essays. What geology was to the Wheeler Survey, these works are to the corpus of Gilbert's writings: a valuable adjunct, a reconnaissance most significant for what it anticipates. But the experience with Wheeler did bring Gilbert and geological science to "an unusual variety of geological features distributed over an immense area," for the greater portion of which "no geological description whatever has been written."[47]

What held Gilbert's own description together was a keen appreciation for the structural geology of the Southwest. His topical essays, like the landscape features he interpreted in them, are related by certain fundamental regularities of geologic structure. Whether he was comparing the Basin Range to the Colorado Plateau, analyzing the adjustment of streams to folds, faults, and variable lithology, commenting on patterns of volcanism, or expounding laws of land sculpture, the foundation of each analysis was a thorough understanding of geologic structure.

A sense of structure similarly affected the organization of his 1875 report. To arrange his data, Gilbert employed, as he termed it, "a systematic approach." Preferring a "description of areas rather than lines" of travel, he abandoned the itinerary format typical of exploration narratives in favor of a topical organization. "Whenever the facts at hand have appeared to warrant a general statement," he declared in 1873, "that has been given in preference to individual facts." Moreover, in rejecting the itinerary form, he was also select-

ing a style of analysis more appropriate for describing existing geologic conditions rather than a format by which one normally recorded earth history. Gilbert's systematic geology did not mean, as
King's did, that he revised the itinerary formula into regional geologic history. Instead nearly all of his important essays and monographs proceed by an arrangement of careful, systematic contrasts,
in which various geologic regions, or systems, or various geologic
processes are compared with respect to their fundamental similarities and differences. Where, in his 1875 report, Gilbert was able
to generate a comparison, the text was innovative and successful;
where he lacked such a structure, the result was little more than a
highly polished set of field notes.[48]

Gilbert's central essay methodically contrasted the "Basin
Range System" with the "Colorado Plateau System"—both of
which Gilbert named. Each region had been examined separately by
other explorers before him—the King Survey had traversed the parallel belts of mountains and troughs that comprise the Basin Range,
while the army engineers had crossed the Colorado Plateau many
times: John Newberry had seen both but had concentrated on the
Colorado Plateau, whose simple stratigraphy could return big divi-
dends for small investments of expedition time. Gilbert explored
the regions well after those pioneers, and it was he who first appreciated their peculiar properties enough to coin distinctive names for
them and to systematically analyze not only their dissimilarities
but also the nature of their unexpected unity.[49]

Both regions, Gilbert observed, had "system," yet each differed
in its manifestation. The Colorado Plateau was an elevated mass of
horizontal strata whose surface rippled with plateaus and mesas,
while the Basin Range (whose strata had been complexly folded at
an earlier period) represented blocks of crust alternately elevated or
dropped along parallel axes. The faults and folds in each case formed
rhythmic belts; their features held consistently over broad regions;
and the primary force acting on both areas was vertical.

Here was an original insight of some power. In the nineteenth
century, the reigning theories of mountain building derived from the
model of a laplacian earth undergoing secular refrigeration—that is,
the progressive cooling of a molten globe. Accordingly, the crust of
the earth was shrinking like the wrinkled skin of a drying apple—so
the analogy went—and mountains were shaped as a result of horizontal compression in the form of crustal shortening. Following the
superb work of the Rogers brothers on the Pennsylvania Survey, the
Appalachian Mountains were enshrined as a paradigmatic example.
Clarence King had extended this model to the Basin Range, inter-

preting the mountains as crests of anticlines and the valleys as syn-
clinal troughs.

Yet Gilbert insisted that the western cordillera had resulted
from vertical forces and that the processes creating the elevated land
masses were still active. This represented a radical change of mind,
for, as he himself admitted, "I entered the field with the expectation
of finding in the ridges of Nevada a like structure (Appalachian), and
it was only with the accumulation of difficulties that I reluctantly
abandoned the idea." To conceive of those isolated, parallel chains of
mountains—like an army of caterpillars marching to Mexico, as
one geologist suggested—as eroded folds was impossible. The analogy of form between the two types of ranges was entirely superficial.
The legitimate analogy was between the mechanical processes
which linked the Basin Range and plateau systems while they differentiated both western systems from the Appalachians. In a carefully crafted passage, Gilbert documented the reason why:

> In the Appalachians corrugation has been produced commonly
> by folding, exceptionally by faulting; in the Basin Ranges,
> commonly by faulting, exceptionally by flexure. The regular
> alternation of curved synclinals and anticlinals is contrasted
> with rigid bodies of inclined strata, bounded by parallel faults.
> The former demand the assumption of great horizontal dimi-
> nution of the space covered by the disturbed strata, and sug-
> gest lateral pressure as the immediate force concerned; the
> latter involve little horizontal diminution, and suggest the ap-
> plication of vertical pressure from below. Almost no eruptive
> rocks occur with the former; massive eruptions and volcanoes
> abound among the latter, and are intimately associated with
> them.[50]

His conclusions were not readily accepted by most geologists;
in fact, hostility varied directly with the distance of the geologist
from the West. It would be twenty years before most European scientists, embroiled in their own national disputes, would admit the
phenomenon. Archibald Geikie, director of the Geological Survey of
Great Britain, was so astonished by the reports pouring in from the
western explorations that he finally toured the area himself—to his
complete satisfaction. And what he announced for fluvial erosion
applied equally to tectonics: "It is unquestionably true that had the
birthplace of geology lain on the west side of the Rocky Mountains,
this controversy would never have arisen." Yet even some of Gilbert's fellow Americans so declined to accept his analysis of the

Basin Range province that he himself returned in 1901 to 1902, 1914, 1916, and again in 1917 for fieldwork toward a classic monograph which, published posthumously in 1928, resoundingly confirmed the insights of his scientific youth.[51]

Volcanics was another theme of some magnitude. At the time of his 1875 report there was "in progress the most active and widespread discussion of the whole subject of volcanicity," Gilbert declared, adding that it was his "purpose to make no theoretical contribution in this place to any of these topics, but merely to present such facts as have come under my observation, with occasionally a brief statement of their relation to the problems of the day." It was an issue he could hardly avoid: the entire Colorado Plateau was rimmed with volcanic phenomena of all types, and the Basin Range was equally the scene of fresh volcanic fields. Directing his observations to "the problems of the day," however, meant that GK selectively edited his field notebooks to test Baron von Richthofen's theory of volcanic sequence, very much as Clarence Dutton would do shortly afterward in *The High Plateaus of Utah*.[52]

This was a borrowed model and, correspondingly, a less distinguished analysis. In keeping with his fundamental motif of geologic structure, Gilbert commented on the different relationships he had observed between volcanics, both intrusive and extrusive, and country rock. On the one hand, vast sheets of extrusive rock had spilled onto the surface without deforming the country rock they had passed through; on the other, such as at Mount Taylor, local deformations in the form of regional doming had accompanied volcanic intrusion. It would take another two years before Gilbert arrived at a theory of volcanism of his own. At that time he would solve the enigmatic volcanism of the plateau system apparent in the Mount Taylor region by scrapping the chemical and petrological formulations of von Richthofen in favor of a physical, mechanical one.

Other Gilbert themes, which would become standard fare in later years, also groped toward definition. The concept of equilibrium was one. Borrowing it from his readings in physics, Gilbert translated this concept into geologic processes until it became a sort of baselevel or ideal condition from which one could measure departures. He applied it, for example, to the profile of a hillslope mantled with lava detritus. Another concept pertained to wavelike rhythms which, to Gilbert's mind, played across the spatial and temporal landscape. In other essays, he made an early attempt to axiomatize erosive processes, formulating two laws of erosion which sought to relate landforms to their underlying structural geology. As an offshoot, he analyzed the process of valley formation in the Colorado

Plateau. Finally, he explored the mechanics of the region's structural geology for clues to the question of crustal rigidity. In 1875, he concluded modestly that "the meaning of these movements of the earth, in vast but limited masses, is, that rigidity is an important factor in the determination of the superficial manifestations of subterranean movements" and that it is necessary to "grant to the rigid masses a depth commensurate with their superficial dimensions, and suppose that the forces which move them are seated still deeper."[53] This was a finely reasoned conclusion—the insight that underwrote his interpretation of structure. Eventually Gilbert made his appreciation of rheology pivotal in assessing competing geological and physical theories of the globe and in helping define his unique hybridization of physics and geology.

Granted the distinctive regional structure of the Basin Range, Gilbert publicized another expression of the Pleistocene which he discovered there. Just as mountain structures had differed in the West from those in the East, so had climate. In the East and Midwest, Pleistocene climate took the form of continental glaciers grinding across the landscape, but in the West it generated pluvials which turned basins into continental seas. The glacials led to the Great Lakes, the pluvials to the huge lakes of the Great Basin. "Those features of the flooding of the Great Salt Lake" which he had observed in 1872 compelled one to admit "it as contemporary with the general glaciation of the northern portion of the continent . . . and to consider it . . . a phenomenon of the Glacial Epoch."[54] Gilbert had discovered a dimension of the Pleistocene to match Agassiz' original discovery of continental glaciation. In the course of a dozen pages, he proceeded to fully outline the core observations and fundamental conclusions which culminated in his 1890 monograph, *Lake Bonneville*.

In his 1875 report, as a spin-off from the Bonneville study, Gilbert commented on the fluctuations in the shore level of its diminutive modern relic, the Great Salt Lake. The lake level had steadily risen since the occupation of its shores by Mormon settlers in the late 1840s. Naturally enough, optimistic agriculturalists had linked the two phenomena genetically so that settlement became the cause for the rising waters in Salt Lake. As the argument ran, cultivation had increased rainfall. Gilbert denied this by an analysis of meteorological principles: "Taking the broadest view, the humidity of the Great Basin depends on air currents, that completely traverse it; and nothing will augment it, that does not either increase the moisture of the incoming air currents, or decrease that of the outgoing." Wheeler had made similar points in his own rather turgid analysis of water supply and agriculture, which he based on an analogy to irrigation in Italy and India. Still, Gilbert admitted that he could not adequately explain the reason for the undeniable rise in lake level. It was not until he supplied the scientific muscle for Powell's *Arid Lands* report in 1878 that he conceived the legitimate cause. Taking a clue from Powell, he guessed that human engineering, particularly overgrazing and the destruction of natural stream regimens, had increased the surface runoff in an otherwise stable climate.[55]

GK's 1875 writing both made for a good report and assured his scientific and exploration credentials. But it lacked precisely the attribute that his great monographs possessed: a complete theme, especially one which approached a mathematical and physical law, one through which he could construct comparative studies. Instead, what he wrote was analogous to the drawings he used as illustrations—a cross section, which was neither a geologic history nor an in-depth geophysical analysis. The report was a perceptive physical geography, a topical travelogue, an anthology of geological essays. It remained for him to integrate his geologic ideas into a comprehensive whole, as he did with his methods of illustration by using the block diagram.

In fact, his summary report for Wheeler was actually fractured into two parts: one for the 1871 to 1872 field seasons, which contained his main contributions, and another for 1873. The latter was obviously composed with great reluctance and at Wheeler's insistence; in it Gilbert satisfied his obligation by writing what was basically a recapitulation of thoughts already formulated and merely extended to a new region. It was a form he might have used on the Ohio Survey, conforming to its "Directions for Collecting and Observing." Yet even here there was a clarity and an excitement to his prose that belied his protests of drudgery. The report did not suggest that Gilbert was arriving at novel, half-articulated insights so much as that he knew more than he committed to the page.

When Clarence Dutton read the 1875 report (so he later wrote to Geikie), he was astonished at a prediction Gilbert had made during one of his few incursions into historical geology. On the basis of a single, ill-preserved fossil, he had dated as Cambrian the block of strata above the great unconformity in the Grand Canyon. There were no equivalent rocks known in the Colorado Plateau region at the time. This, Dutton wrote, "was a little too much so Powell and myself dissented from Gilbert." When events later proved Gilbert correct, Dutton exclaimed: "I cannot imagine how Gilbert could have had the courage to adopt such a preposterous conclusion, and yet be right about it after all."[56] There were many other seemingly preposterous observations lodged in the report, based, so it appeared to his contemporaries, on data no less questionable. When Gilbert finally pursued them in depth, his conclusions surprised a good many geologists.

The Lieutenant and the Major

During his final year with the Wheeler Survey in Washington, Gilbert made friendships, established social habits, and entered into the life of scientific societies in patterns that endured for decades; for the remainder of his life, he spent at least half of each year in Washington. During this turbulent period, too, he began his decisive friendship with John Wesley Powell, divorced himself from the frustrations of the Wheeler Survey, and married. The transformation from the Ohio Survey to the Wheeler Survey had advanced him from the status of amateur naturalist to accredited exploration geologist; the events of the winter of 1874–75 were to propel him still further toward the life of a scientific administrator, an internationally recognized theoretician, and an active member of Washington's scientific society.

During the long months that he wrote reports, Gilbert and Archibald Marvine became exceptional friends. Evidently their companionship proved a bit too robust; heaving a ball once "near the market" only brought the police. Along with his old friend Howell, he took dancing lessons, while a triumvirate consisting of Gilbert, Howell, and Henry Henshaw, masterminded by Gilbert, established some unusual living arrangements. They selected a boardinghouse on a street between 6th and 7th for their meals, then searched for a rooming house convenient to the survey offices about two miles away. Henshaw recalled the adventure thus: "Gilbert as the originator of the scheme was appointed a committee of one to visit all the nicer and more pretentious dwellings in the neighborhood, none of which were rooming or boarding houses," and ask if rooms could be had. "The astonishment, inclining in one or two cases to something like indignation, exhibited by several of the matrons in turn when thus approached by a stranger, without letters of introduction, formed the subject of successive reports of progress to the committee of the whole and furnished hilarity for several days." Yet the calm demeanor which had scrutinized Venus from the depths of the Grand Canyon was hardly to be routed by such rough water. "It speaks volumes for Gilbert's appearance, poise, and diplomacy," Henshaw noted with some amazement, "that not a single absolute refusal was met with." The plan worked, with the incidental benefit of keeping the trio "in good trim. Rain or shine," Henshaw remembered, "we made the trip there and back twice a day." Their appetites were naturally "in proportion to the length of their walk."[57]

Not that they failed to exercise. There were frequent tramps, and even at that time the Washington countryside remained "almost in its primitive state." With the Potomac nearby, GK had abundant opportunity to boat. Usually he sought out some companion, although formal society was anathema to him. In the evening he occasionally attended the theater or lectures, but more often he preferred playing cards (especially euchre and whist) or reading aloud from favorite books for the amusement of small gatherings of friends.

Those friends commonly belonged to the scientific community of Washington. Since he had joined the Wheeler Survey, Gilbert had spent at least a portion of his winters in Washington and had discovered the scientific societies available there. He participated actively in the Philosophical Society of Washington, founded in 1871, which greatly amplified his scientific associations, and in the Geological Society of Washington, created in 1873 with Gilbert as one of its chief organizers and first secretary. Here he discovered a forum for the ideas he later developed into major monographs, as well as a comfortable society of friends. Increasingly, the two became synonymous. In fact, more and more of his social affairs developed solely around the dynamic personality of Major Powell. Together with Henshaw, he and the Major would avidly play cards once a week, terminating the hands at 10:30 so the party could adjourn to cigars and light talk for another hour or so. From Powell's example, Gilbert took up billiards, using it to replace an earlier interest in chess, and "in this," Henshaw noted, "as in all other things he elected to do, he soon became surprisingly proficient."[58] Clearly GK was attracted to games which not only engaged his friends but involved almost mathematical concentration, especially in the form of counting (as in cards) or geometry (as in billiards). Probably, too, there were lively exchanges of riddles and "problems."

His writings for Wheeler, however, languished. His description of the 1871 and 1872 field seasons went briskly, but the report for 1873 advanced only with constant prodding from Wheeler and General Humphreys. And, for all its assets, his 1875 report, when finally published, left Gilbert deeply dissatisfied. In his progress report for 1871, he had written that "the most we can claim to have accomplished is a reconnaissance of our field" but that "the achievement of more thorough work in connection with exploration is neither possible nor, in every sense, desirable." By 1874 he thought otherwise. He expressed his frustration when he wrote of the Amargosa Mountains: "It was extremely tantalizing to see there not less than 8,000 feet of bedded rocks so beautifully displayed, and yet be unable to examine a single stratum."[59]

Even though the survey's premier paleontologist, F. B. Meek, had "immortalized" him "on a bryozoan," Gilbert's discontent grew. In a typewritten preface he inserted in those copies of his final report which he personally distributed, he wrote with a tinge of resentment that

> to study the structure of a region under such circumstances was to read a book while its pages were quickly turned by another, and the result was a larger collection of impressions than of facts. That many of these impressions should be erroneous was inevitable, and no one can be more conscious than I of the fallibility of what I have written.[60]

Meanwhile Wheeler—under orders from the wary General Humphreys—had "negatived" Gilbert's desire to place some articles in the *American Journal of Science*, if they would appear prior to the official publication of the 1875 report. He also denied Gilbert's wishes to cite evidence derived from communications with "rival" surveys—that is, Powell's. "I have found it necessary in most cases to erase these," he wrote. Moreover, at this time affairs in Gilbert's personal life were changing in ways that rubbed against the Wheeler Survey style, and such arbitrary editorial acts (as Gilbert interpreted them) only increased the friction. Besides, failure to document his sources could damage his professional reputation. The army eventually recanted, but by then Gilbert had resigned. By then, too, his actions had indirectly helped (though with generous assistance from Wheeler and Humphreys) to discredit the Wheeler Survey.[61]

Developments such as these led Lt. George Montague Wheeler to spend his final years in a bitterness that bordered on outrage. In 1879 the U.S. Geographical Survey West of the 100th Meridian was dissolved, its functions absorbed—along with those of the other western surveys—into a new civilian bureau, the U.S. Geological Survey, with Clarence King as director. Wheeler, like F. V. Hayden, was allowed to complete the publication of his atlases and reports. In 1881, he was appointed U.S. commissioner to the Third International Geographic Congress and Exhibition at Venice. By 1883, however, his health along with his ambition was broken. As he reported to an army review board, "The field trips were often attended by the greatest hardship, deprivation, exposure, and fatigue, in varying and often unhealthy climates . . ." The board voted retirement at his own request, declaring him incapacitated for active service. Although Wheeler retired officially in 1888, he continued to write his reports until the last was published in 1889, and by an act of Congress in 1890 he was awarded the rank and pay of major.[62]

Wheeler had identified his career with the army. Whenever he published a map showing the routes taken by his topographers, it also included the routes of all previous army engineer explorers in that region. When he presented a roster of personnel, he included special remarks on each of his officers, while relegating his civilian scientists to a list like that they might use to catalog new species of birdlife discovered during the expedition. Obviously he felt they should be grateful to the army rather than vice versa, especially as the volume of applicants—many of them graduates of new scientific schools—for those few scientific posts swelled with each passing year.

Backed by General Humphreys, Wheeler felt confident that the army would triumph in its competition with civilian surveys because it justified its program on that crucial necessity—the topographic map. In his report on the international congress, he wrote long panegyrics about the multiple values of such maps and invariably concluded that the military alone should execute them. "By natural selection," he admonished his critics, such work inevitably devolved to the military. Yet he was puzzled and outraged to discover that in the United States "filibusters . . . under the flag of the sister sciences, march forth to the subjugation of this, in common with all other scientific knowledge, in order that it may be finally prostituted to their sole ambitious aims." He sneered at civilian topographic maps which were unsuited to military purposes, while the civilians reversed the charge to claim that Wheeler's atlas was worthless for geologic purposes. In the end, the important facts were two: the army was in disfavor, and the civilians had coopted and upgraded the better methods of the army way. Ironically, as the Wheeler Survey became more efficient, it also became more dependent on civilians. It thus lost its justification as a military venture. Congress concluded that the important work was done by its civilian employees and that army administration, like army escorts, was only a costly ornament.[63]

The survey, like Wheeler himself, represented an almost grotesque struggle to achieve a great romantic gesture. Where King unveiled the Diamond Hoax and made thrilling ascents in the High Sierras, Wheeler could only belatedly send parties to the long-familiar Yosemite Valley and Lake Tahoe. Where Hayden could publicize the Yellowstone and agitate for its park status, Wheeler dragged his men across Death Valley, generating local criticism and several cases of heatstroke. Where Powell leaped to public prominence by traversing the dramatic canyons of the Colorado River, Wheeler reversed the procedure. But traveling upstream reversed the consequences as well, yielding consternation and ridicule rather than glory. Here, indeed, is the symbol of the Wheeler Survey: conceptually and institutionally, no less than geographically, it traveled against the mainstream. With great labor and iron purpose, it achieved numerous valuable results, yet it disintegrated. The Humboldtean cosmos in which it had been conceived had exhausted its intellectual vigor by 1871, and lacking that scientific energy the Wheeler Survey became the simple reenactment of tradition as, at the Grand Canyon, it rendezvoused with the memory of Ives and left without advancing the scientific or military significance of the canyon. Wheeler's work stood to Powell's exploration as his limp, florid prose did to Powell's crisp commands. Even when Wheeler published, his text was often sadly out of date.[64]

As a scientific enterprise, the survey was antiquated—trying to stuff vast geographical information into a Humboldtean design. The survey still thought of its scientists in terms of the all-purpose physician-naturalist. Yet, just as the army surveys were eclipsed after the Civil War, so, figuratively speaking, was Humboldt after Darwin. There was no place in the army for that new science. The only place open to its old style of exploration was Alaska and the poles, where the army eventually went. The closest Wheeler came to a Mount Chimborazo, on whose slopes Humboldt had captivated world imagination, was Sierra Blanca in Arizona. Climaxing an arduous climb, Wheeler eyed the panorama and declared himself satisfied that no white man had seen the sight before. Probably none had, but it was a long time before anyone bothered to see it again. Instead of quality or even priority, Wheeler argued increasingly by quantity—by the number of maps that rolled off the presses like the miles beneath his odometer.[65]

Hence, Gilbert's frustration. But he should have been more magnanimous. The survey had tempered him into a highly competent field geologist; its reconnaissance style had forced him to organize his data into broad geological themes; it had compelled him to experiment with modes of geologic explanation and literary composition; and it had familiarized him with regions and subjects he later hammered into his major monographs. Except for its brief excursion into stratigraphy, the text of his final report reads like a table of contents to his subsequent career in the West. In four years with Wheeler he staked out geologic claims which he would later mine.

Yet GK's experiences during those four years stood in marked contrast to what he saw of Powell. Consequently, the geologic and literary training he got from Wheeler, the observations he indexed in his rich field notebooks, the crisp insights he epigrammatically phrased in his reports, and the volumes of additional reading he pored over at night—all were eventually put to the service of Powell. Wheeler's program had been ambitious, incorporating in his maps a system of land classification, Indian ethnography, and irrigation studies which, by 1874, largely replaced geologic studies. Yet it was all preempted by the civilians, especially Powell, just as the Major had shanghaied the young lieutenant's chief geologist. Indeed, Powell's own Civil War experiences had taught him what he needed to learn from the army, and his own obsession with topographic maps probably stemmed from his years as an officer under Grant. Still, the lessons were not completely absorbed, and Gilbert's timely transfer of surveys meant a valuable transfer of acquired skills. When Powell sent him to his Kanab base line in 1878, Gilbert found the work generally bungled: he had to completely resurvey the topography. In order to train Powell's men, he handed them copies of a general textbook on surveying and the reports written by Lieutenant Marshall of the Wheeler Survey.[66]

What most attracted Gilbert to Powell was probably what Powell had not absorbed from the military—his openness, sociability, and enthusiastic vision. There was a promise of fewer orders and more science. GK was never happy too far and too long away from the Nutshell, and the social circle collecting around Powell, like that around Newberry, promised to emulate it more than did the military chain of command, which seemed to elicit more sympathy from Gilbert for his mules than for his officers. In short, what hurt Wheeler most from Gilbert's standpoint was probably the same thing that damaged him before the Townsend Committee in 1874. The committee commended "all the surveys for their excellent work" but censured Wheeler (and Hayden) for bad manners.[67] Even before he formally transferred, Gilbert had clearly entered into the scientific collaboration with Powell that continued for nearly thirty years.

In his letter of resignation to Wheeler, Gilbert hinted that he would like to be relieved from report writing and asked—without giving reasons—for a month's leave of absence until the resignation would take effect. Stunned, Wheeler denied both requests.[68] The explanation again derived from Gilbert's growing ties to Major Powell, though in this instance Powell only engineered the general circumstances. On January 10, 1874, at a dance in the Major's house, GK met Fannie Porter—the sister of Archibald Marvine's wife. Walks followed the dance, then visits. That summer, during Gilbert's month-long tour of New England to attend the American Association for the Advancement of Science meeting in Hartford and his sister's wedding, there were longer visits. His request for leave had been for this month—a leave of absence which he wangled anyway. After his tour of duty with Wheeler was concluded and his reports were sent to the printer, Gilbert married Fannie Porter on November 10, 1874. He was thirty-one years old.

Two weeks later he wrote Powell from the Nutshell:

> Fannie and I succeeded in getting married at the time appointed (and have carried out what programme we had for after performances to our satisfaction—to date I am not yet tired of being married but) I am getting a little anxious to be at work—partly because it has come to be more natural to me than play, and partly because I ought to be earning something. So I am going to Washington in a few days with the intention—if you have not changed your mind—to begin work with you at once.[69]

The Major had not changed his mind. Indeed, he bristled with plans for the coming field season—plans which would, over a period of years, throw the trajectory of Gilbert's career into a wide orbit around his own and which, in the short run, would fulfill Gilbert's longing to once more be "astride the occidental mule."

3. The Major Years

And here, oh, what a view is before us! A vision of glory! Peaks of lava all around below us. The Vermilion Cliffs to the north, with their splendor of colors; the Pine Valley Mountain to the northwest, clothed in mellow, perspective haze; unnamed mountains to the southwest, towering over canyons bottomless to my peering gaze, like chasms to the nadir hell; and away beyond, the San Francisco Mountains, lifting their black heads into the heavens.

—John Wesley Powell

The laccolites were there for the seeing.

—Grove Karl Gilbert

"The western fever"

When he joined the Powell Survey in 1875, Gilbert found it undergoing a metamorphosis similar to that which the Wheeler Survey experienced about the same time. The survey had begun in 1869 with a spectacle—Powell's two traverses down the Colorado River by boat. In 1871, with congressional funding, the U.S. Geological and Geographical Survey of the Rocky Mountain Region pitched its tents at Kanab, Utah, for several seasons of exploring and mapping. Organized rather along the lines of the river trips, the field parties consisted of relatives and acquaintances of the Major. His brother-in-law, A. H. Thompson, supervised the fieldwork, as he had most of the second river voyage—in both cases while Powell lobbied in Washington. With help from Mormon frontiersmen, the survey enjoyed some successes—Thompson, for example, located a route to the Henry Mountains and the Escalante River, the last range and river to be discovered in the United States. After they completed a map in 1873, Thompson assessed his operation with some pride—

"we done middling for greenhorns," he wrote Powell. But Wallace Stegner has more appropriately characterized their labors as "Major Powell's Amateur Hour." In fact, Thompson was forced to rely heavily on local talent, especially prospectors, for guides and packers. "Our boys," he confessed in 1872, "don't amount to a row of pins in handling animals." That was true for more than livestock.[1]

Powell revamped the survey. In 1875, he published his narrative for the 1869 river voyage and added his thoughts on the geology of the river; together they comprised *The Exploration of the Colorado River of the West.* He kept his seasoned field hands and brought in two geologists—Gilbert from the Wheeler Survey and Capt. Clarence Dutton on loan from the War Department. In sum, in 1875, while Newberry's report for the Macomb Expedition was in the hands of the public printer, as the Hayden Survey left Colorado in search of new spectacles, and as Wheeler sprawled about the western landscape, the Survey of the Rocky Mountain Region sent three field parties to the Colorado Plateau: one each under Thompson, Dutton, and Gilbert.

Actually, GK's first chore for the Major was to arrange his fossils. In a sense, this bit of scientific housekeeping characterizes much of his relationship to Powell. It is unlikely that Gilbert minded much for, between his recent marriage and the prospects of his new job, he was quite exuberant. At the same time, he helped Thompson in reducing triangulation notes to a map, worked with Charles White on the fossil report, and spent many evenings as a nurse to his friend and brother-in-law, Archibald Marvine, whose health had been broken during a field season in the Colorado Rockies. But it was a bustle not everyone appreciated. William Dall wrote F. B. Meek:

> We have been a good deal annoyed since I returned by the ungentlemanly behavior of a person named Gilbert—now in Powell's service. On the pretense of working on Powell's fossils he made a perfect bear garden of the room, without making any sign of apology or excuse and wanted Prof. Baird to open your library cases in order that he might use your books for study. Of course Prof. Baird declined to do any such thing and for a time at least we are rid of this nuisance.[2]

He had better luck in the field. After a year in Washington, "the western fever," as he called it, of which he had a terminal case, raged. He returned gladly to southern Utah and northern Arizona, both areas familiar to him from his years with Wheeler. There was less adventure but more science, and his years with the Wheeler Survey had taught him well how to plan for a season's fieldwork. His friend Howell joined him much of the time, and his routes periodically intersected those of Jack Hillers, the survey photographer, and, of course, Dutton. There was a change, too, in that, cognizant of the basic geologic structure of the plateaus, he addressed questions of land sculpture, or physiography, the topic which had aroused his curiosity in the Maumee Valley. Powell undoubtedly reinforced this tendency, and together they explored streams whose relationships to geologic structure were far more striking than those around Lake Erie.

From Salt Lake City in late June, the parties traveled south along the Wasatch Mountains. When they reached the High Plateaus, the three split up—Thompson continuing topographic mapping, Dutton touring the volcanic scenery of the plateaus, while Gilbert turned east to explore the magnificent monocline that formed their eastern border. Waterpocket Fold was then, as it still is today, wild country—vast walls of castellated cliffs carved out of the chromatic rock layers thrown up obliquely by the great warp. The land is arid, and passages through the cliffs are possible only in breaks created by rivers whose headwaters belong in the plateaus to the west. The streams breach the fold in ways as dramatic as the incision of the Green River through the Uinta Mountains, a scene that inspired one of Powell's genuine geologic insights.

Travel was rough. At times there was no opportunity for even the luxury of a tent. Gilbert recorded one such episode where the whole party—Graves, Bell, Sorenson, and himself—slept out. They didn't reach their destination "until 1:30 p.m. and left at 5 p.m. Darkness overtook us and we barely made a waterpocket in the descent when we were forced by the uncertainty of the way and by weariness to stop." As if that was not enough, they started out at 4:15 that morning "and reached breakfast at 6:30." Yet Gilbert enjoyed it, lavishing his attention between the geologic displays and his mule, "Lazarus, Duke of York."[3]

En route along the Waterpocket Fold, Gilbert could see the dark peaks of the Henry Mountains to the east. He studied them frequently from his distant perch. By the time he arrived at the Henrys on August 23, he had formulated a clear conception of their structure. He wrote, "My ideas of yesterday in regard to H.V. are confirmed by this view. It is in bubble form or tumor form, the strata being nearly level on top and the crusts controlled by dikes which are radiate in form." GK had conceived the structure of the mountains before he ever actually visited the scene. By the following day,

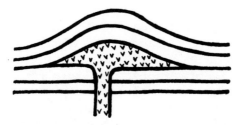

Gilbert's diagram of a laccolith, August 22, 1875

he had an explanation for the erosional processes operating on the scene as he wrote *"A General Note on Terraces"*—which he conceived as "not built of *debris* but of rock *in situ* capped by debris."[4]

For a little more than a week he explored the Henrys, eyeing all of them. But his crucial insight is contained in a small diagram which he drew the day before he rode Lazarus to Mount Ellen. It summarized the thoughts he later elaborated into his monograph on the Henrys—the structural analysis of a novel process of mountain building whose form he called the laccolite. The accompanying text reads:

> The types of Hillers and H.V. are somewhat different. The radial dikes of the latter are doubly represented by the trim radials of the former. The concentric dikes of Hillers do not appear in H.V. If Hillers be an extension of the H.V. type, then only the Trias was lifted and the carboniferous either lay below the seat of action or below a distributing reservoir.[5]

The party moved north to the San Rafael Swell and the Dirty Devil River, amid a strange landscape which at last conformed to the romanticized drawings of the Ives report. "Egloffstein took his canyon topography," Gilbert remarked, "from such a piece of drainage as lies below us in the clay flat with its ramifying arroyos."[6] At Capitol Reef, where the Dirty Devil breached Waterpocket Fold, they turned west. By September 10 the field parties were breaking up—the packers returning to their southern Utah homes, the geologists moving north to Salt Lake City and, eventually, to winter assignments in Washington.

The 1876 field season began routinely, with visits to Rochester and Jackson, Michigan, en route west. "At Chicago," Gilbert wrote on August 11, "I breakfast in the Fremont House Restaurant with Whitfield and then in the depot meet Hillers. At Aurora Thompson joins us." A month later, with field parties divided as before, circumstances were less pleasant: "A chapter of accidents. Frank kicked by Little Nephi in the shin. Lightfoot about played out and down twice. My pack bucked off and 3 alforjas torn. Evening spent in repairs. Water in pockets is bad." At the time Gilbert was again approaching the Henry Mountains for an extended study, for the previous fall he had written Powell that the Henrys were "a group of structure that will well repay a month's study." Early in September 1876, he established camp for that study.[7]

The geology progressed rapidly. Most of Gilbert's time, in fact, was spent in topographic labors which he thoroughly relished. His work proceeded in a stylized way, developing along two approaches to the landscape around him: the panorama drawn from a mountain summit plus a special investigation into topics that appeared problematic or curious from the perspective of the peak. The work was routine, organized in a systematic fashion he probably learned from Wheeler. Most of his notes consist of finely wrought drawings of the mountains. There was little hesitation about his survey. As he later wrote in the preface to his monograph on the Henrys: "So thorough was the display and so satisfactory the examination, that in preparing my report I have felt less than ever before the desire to revisit the field and prove my conclusions by more extended observation."[8] The reason is simple: GK knew what he wanted to look for.

On October 11, after an evening shower, the camp awoke to find "snow on Mt Aquarius and Hillgard." Rain continued. On the twelfth Gilbert wrote: "Work was stopped yesterday by rain which began about 4 PM. There has been some intermission since, but this morning we are at the lower edge of a great cloud and a drizzle of rain half the time. Two grand showers yesterday moved from Aquarius Plateau across the Waterpocket fold and the Blue Gate." Stymied from the usual round of surveying, he organized his thoughts on the structural meaning of the Henrys over the course of some twenty-seven pages in his notebook. Systematically addressing each aspect of the problem, the notes present the entire theory of the laccolith almost exactly as it appeared in his 1877 monograph.[9]

It is evident that Gilbert's enthusiasm for topographic labor derived from the same impulse that led him to map the Cohoes potholes: it allowed him to make the measurements he needed to verify his theory, in this case, on laccolith formation. The long notes written on October 12 began with a calculation of "spherical excesses," figuring the volume of the igneous cores to the mountains. Evidently he imagined the laccolith shape as a spherical triangle or, indeed, as a round loaf of bread.

The rain continued for several more days: "Water in the frying pan indicates ¼ inch rainfall since yesterday." Gilbert outlined his monograph and, perhaps inspired by the rain, collected his thoughts on the erosional processes of the Henrys. Meanwhile the "boys returned from Rabbit Valley with supplies and letters." The weather worsened—"work almost barred by the snow storm and a gale of wind . . . level frozen. Theodolite levelled by guess and the horizon." Things were little better in camp: "Yesterday level rolled over once on a hill and smashed my looking glass. Today the cook's fire ran through the oak leaves and burned the saddle, saddle blanket, spurs, overcoat and saddle bags—all a little so that repairs are needed."[10]

Most of Gilbert's speculations concerned erosional forms. Well understanding the meaning of the broad stream terraces flaring off the mountain flanks, he added to that understanding some mechanisms to account for the "rounding of the crests of Utah badland ridges" as well as the observation that "the orders of size and ruggedness of the mountains are reciprocal"—"purely a matter of climate as I conceive."[11]

In the process of surveying, Gilbert named the major landforms. Most of the names belonged to survey members or to those geologists whose works were among his "chief sources of information in regard to igneous mountains in general." Significantly, he refused to name anything after himself. He added wryly, however, that "if any of these gentlemen feel offended that their names have been attached to natural features so insignificant, I can assure them that the affront will never be repeated by the future denizens of the region. The herders who build their hut at the base of the Newberry Arch are sure to call it 'the Cedar Knoll'; the Jukes Butte will be dubbed 'Pilot Knob', and the Scrope, 'Rocky Point'."[12]

By mid November, Gilbert concluded his fieldwork with a final tally of mule accidents: "Panguich rolled over today into Curtis Creek. This is her third roll on the trip. Beck has accomplished two and Gomas, Joel, and Louey one each. Our little train of 9 animals has attained to 7 [actually 8] rolling scrapes." On the other side of the plateau, Dutton was wrestling with the same conditions. He wrote Powell: "I have also solved another interesting problem viz: how high steep and rough a hill a mule can roll down without getting killed."[13]

As he neared the end of his stay, Gilbert wrote in his field notebook: "Sunset. 5:35 PM. The Henry Mountains cast their shadow against the La Sal."[14] When his report was finally published in 1877, the shadow of his laccoliths touched a great many other mountains in the Colorado Plateau and made the Henrys known to more than herders. On November 22, he reached Salt Lake City and boarded a train for Washington.

Gilbert's remaining fieldwork for the Powell Survey brought him to themes and areas he had traversed with Wheeler. In 1877 Powell directed his forces into irrigation studies not much different from those Wheeler had grappled with. Consequently, GK launched a program of reading and instruction on topics related to irrigation; most of the literature was written by army engineers. He was already well versed in the mechanics of stream behavior and, from his Ohio days, in the problems of fluctuations in lake levels. As he wrote Powell in July, "The gauging of streams, the study of beaches, and the study of recent faults go well together and make the great part of my fieldwork." In addition, he calculated acreages of arable and irrigable lands, which he derived from a general axiom that "to produce a rich farming district in Utah, there are needed high mountains and low valleys in juxtaposition"—a fact Wheeler had buried amid a perfect rubble of complicated syntax.[15]

Perhaps more important, Gilbert cultivated a concern with instruments. It was a measure of the engineer in him, a manifestation of the same literalness that made him visualize local warpings in the Great Lakes as a series of dams. In Utah he developed some ingenious techniques for measuring the oscillations of the Great Salt Lake and drew up the blueprints for a new surveying instrument. At the same time, bothered by inaccuracies in determining altitudes by the use of the barometer, he gathered notes which eventually led to a novel hypsometric technique and a new formula to supplant that of Laplace. The problem of measurement errors continued to press him, so much so that, at times, the problem of instrumentation itself became a puzzle almost as engrossing as the current geologic riddle. Where most other geologists scrutinized fossils, Gilbert examined his instruments; where they envisioned the geologic map as a means of synthesizing geologic history, Gilbert saw its value as measurement, a blueprint of the landscape.

Not that all his ventures with instruments proved successful. At meetings of the Philosophical Society of Washington, he learned to his disappointment that the special graph he had devised to calculate hypsometric values was less original than he had believed. The story was similar for a surveying instrument he designed. He assigned his topographer, Willard Johnson, to construct a working model. But, as Johnson wrote to a friend: "Now I learn that during the war more than thirty applications for patent were made for a device embodying precisely the same principle, and varying only slightly in detail, and that the idea is probably as old as Archimedes." [16] Little matter—it is the impulse to confront the landscape through such instruments that is important, and it was by the application of another principle from Archimedes that Gilbert constructed his theoretical model of laccolith formation, with consequences far outstripping those of any improved theodolite that might have found its way to the Patent Office.

He had the chance to demonstrate that interest in numbers with the 1878 field season. Opening the campaign with eleven men, thirteen packs, and twenty-six animals on August 5, he guided his crew south to Kanab. Here he assumed A. H. Thompson's old role as field marshal for Powell's topographic efforts. The crew attempted to bring the mapping quality of the southern portion of the state into parity with that of the north, executed by King and Wheeler. But Gilbert found the old triangulation network so inaccurate that he began a resurvey of the entire region. Directing the operation left him little time for much else. He assured Powell, however, that "we are doing work and will bring out a good chain of triangles from base to base. We have been somewhat delayed by rain and have run out of provisions, but otherwise all has gone well." [17]

He toured northern Arizona again, this time crossing the Kaibab Plateau to the north rim of the Grand Canyon, then sweeping around the canyon in a great circle from Kanab to the San Francisco Peaks and northwest to Saint George, Utah. There were some surprises: "For the first time in my life I have seen a snake drink," he wrote. But, for the most part, "I have seen little new in geology." Practically the only remarks of note were a classification of terraces and the recognition that the terraces terminating the High Plateaus to the south did not result from "planation" but depended on "the rock structure including bedding and displacement." [18]

During this time, most of his efforts went to train and supervise his field parties. Near Moenkopi, Arizona, this resulted in some anxiety about one of his men, named Wheeler: "I am afraid he is in trouble. He let go one of the men I sent with him and hired in his place a man whom we know to be a horsethief just escaped from the sheriff. I have sent an Oraibe Indian to warn him, but fear he will be too late. I start myself in the morning." But there were few problems with Wheeler, who "lost no horses by the thief in his employ but on the contrary liked him exceedingly and was much annoyed by my letter of warning. The man was arrested in Moenkopi and escaped a few days afterward." Gilbert ended the season with a full muster of men and mules on December 18. [19]

Clearly GK was still arranging Powell's fossils for him. It was tedious labor, which he nevertheless performed conscientiously. He made few complaints—he rarely did when he had the opportunity for fieldwork, even when that tour of duty exposed him to areas of comparative familiarity. And, as with most of his chores as a field manager, he freed Powell for political maneuverings in Washington. In this case Gilbert's was a silent contribution to an important enterprise that ever after dictated much of the shape of his scientific career. From a tent in Kanab, he addressed a letter to Powell on that crucial subject: "I am anxious to hear all about the Nat. Acad. and the consolidation of the surveys."[20]

By Virtue of Its Ensemble: Powell, Dutton, Gilbert

For all its often slapdash organization, the Powell Survey catalyzed a remarkable scientific collaboration, mindful of Hutton and Playfair, Chamberlin and Moulton. Dutton perhaps put it best. "With a bond of affection and mutual confidence which made this study in a peculiar sense a labor of love," he wrote of the intellectual alchemy between him, Powell, and Gilbert, "this geological wonderland was the never-ending theme of discussion; all observations and experiences were commonstock, and ideas were interchanged, amplified, and developed by mutual criticism and suggestion. The extent of my indebtedness to them I do not know. Neither do they." Between them the three men not only brought substantial portions of the West into geographic knowledge but left an innovative legacy of landform studies which led, largely in the hands of William Morris Davis, to the new geology of land sculpture. This, in turn, became the cornerstone of the American school which matured in the early twentieth century. In popularizing the Colorado Plateau, moreover, they gave American geology the greatest symbol of its heroic period, the Grand Canyon. The canyon, Dutton proclaimed, was "the sublimest thing on earth. It is so not alone by virtue of its magnitude, but by virtue of the whole—its *ensemble*." The same might well be said for the Powell Survey.[21]

The three not only gave geological descriptions to previously unexplored terrain but also deliberately sought to relate geological science to other branches of science and culture. In this ambition they followed the example of Dana and Newberry, who, via laws of evolutionary progress, had joined geology to biological and theological thought. It was Powell's achievement to show where the broader questions of the earth coincided, by means of similar evolutionary laws, with social and philosophical schemes. In practical terms, he endeavored to adapt American social, political, and economic institutions to the environmental conditions of the West. It was Dutton's role to join geology to the larger questions of chemistry and geophysics. By assuming that evolution affected every geological phenomenon, he could relate the development of landforms, the petrological sequence of volcanic extrusions, and the history of the earth. His practical contribution was to adapt aesthetic theory, especially that common to architecture and painting, to an appreciation of the new western scene. Gilbert's peculiar achievement was to pioneer in the effort to unify geology by analogy to mechanics and mathematics. His indifference to the earth's historical development segregated him somewhat from the mainstream of earth scientists. His practical ambition in the West was to adapt scientific engineering to it. Significantly, his major monograph on the Colorado Plateau, unlike Powell's and Dutton's, did not deal with the Grand Canyon.

In sum, while all three men found the structural geology of the plateau an inexhaustible topic of investigation, Powell approached it by analogy to paleontology, Dutton by analogy to chemistry and architecture, and Gilbert by analogy to mechanics and civil engineering. As a team, what they lost in depth or intensity of study they gained in breadth and comprehensiveness. After a hundred years, the larger interpretation of the plateau country is still the one which Powell, Dutton, and Gilbert collectively gave it.

Of Gilbert's friendship with John Wesley Powell, William Morris Davis has written that "it was the greatest determining factor of his mature life." The two men were like brothers, though it would be difficult to find two men of such different temperaments. Gilbert was a classicist and a scholar, a man who abhorred controversy; Powell was a romantic, a reformer, and a scientific soldier of fortune who had lost one arm at Shiloh and risked the other in the trench warfare which the Gilded Age witnessed annually in Congress. Where Powell was intuitive and didactic, Gilbert was systematic yet devoid of both pedantry and preachiness. Whereas Powell tended to present his insights as a single stroke of discovery, reaching boldly out to the new and unexplored, Gilbert's speculations arose from methodical comparisons and contrasts, carefully moving into new ideas through analogies to old ones. A strange companionship but, as all their contemporaries remarked, the men were ideal complements and foils to each other. What both shared was a marvelous talent for abstraction, for distilling a scene into its essential lessons. Their friendship was as deep as either knew.[22]

The son of an English immigrant, Methodist minister, and outspoken abolitionist, Powell absorbed from his father a talent for didactic preaching and an evangelical zeal for reformist causes. His formal education was sparse, growing from local resources in rural Ohio and Illinois; it was biased toward subjects in natural history. Much of it was self-education, though he enrolled briefly at Oberlin, Illinois College, and Wheaton. His early jobs included teaching in country schools and organizing collections of specimens as well as field trips to search for more. As if in preparation for his later career as a western explorer, he even piloted a skiff down the Mississippi River.

The formative event in Powell's career was the Civil War: its lessons stayed with him no less than did the tortured stump of his amputated right arm. The war taught him the meaning of organization, and it fired the ambitions of a rural Illinois youth by showing how, through glorious exploits on the fields of battle, one could rise to positions of command. While he detested the slaughter of war itself, he relished its moral equivalents. He left with the rank of major—a title he never surrendered. The war also brought him to the attention of U. S. Grant, and it was through Grant's intercession that he was allowed to draw on army rations for his early postwar expeditions west. A few years later it brought him that versatile captain of ordnance, Clarence Dutton, whose geologic work for Powell was funded by the War Department. The army experience even left its imprint on his prose. His dictations read like a series of commands, as though he were exhorting troops; they were charged and dramatic, as though issued in the midst of great battles. It may be, as well, that Powell's fascination with maps stemmed from the utility they demonstrated in war campaigns.

In 1869 he discovered a way of blending his youthful explorations with the heightened sensations of war. This was, of course, his epic traverse of the then unknown canyons of the Colorado River. The journey was crucial for Powell as both an intellectual and a reformer: it catapulted him to national prominence, almost as a kind of war hero in the battle with nature at its wildest; it designated his favorite fields of study; and it vividly suggested his important geologic themes, particularly the relationship between erosion and structure. He established his scientific reputation with the publication of two reports: *The Exploration of the Colorado River of the West* in 1875 and *The Geology of the Uinta Mountains* in 1876.

The reports merit analysis, both because of Powell's role as director of an important western survey and, later, of the U.S. Geological Survey and because of his close ties to Gilbert. They show an imagination which relied heavily on morphology and evolution as organizing principles. That is, Powell extended paleontological methods to new fields of geology—in particular, to the study of land surfaces. "Geology," he wrote, "is expressed in topography." In other words, the surface shapes tell one about the historical dynamics of the landscape, just as fossils reflect the evolutionary dynamics of organisms. The topographic map was the equivalent to a structural skeleton; like paleontological remains, it represented an epoch of time. It was essential to realize, Powell lectured, that "a mountain was more than a mountain, it was a fragment of earth's history." By elaborating the method of paleontology, as organized by evolution, he was able to incorporate new phenomena into the canon of geological science and establish a precedent in the methodical search for similar fossil equivalents that could provide information about the historical evolution of the earth. Interestingly, even in his imagery there remained a blend of organicism and religious allusion reminiscent of Newberry. He spoke of the "embryology and growth of mountains" and hoped that someday "there would remain the writings of those inspired with the knowledge that a mountain has structure, that every hill has an appointed place and every river runs in a channel foreordained by earth's evolution." Through his reports, combined with his lecturing skills, Powell was able to bring his new geology into parity with the old. In Gilbert's words, "Whereas geologic history was formerly read in the rocks alone, it is now read not only in the rocks but in the forms of the land and the arrangement of streams." Eventually, as William Goetzmann has observed, "uniformitarianism itself seemed almost his own invention."[23]

It is not difficult to imagine Powell as a romantic naturalist in the vogue of Newberry. With Powell the evolution of the physician-explorer into the professor-explorer was complete. Moreover, that mix of poetry and scientific order which Newberry saw for the fossils embedded in coal Powell perceived in the fossil remains of geologic structure—the surface landscape. "He who can see a mountain range, or a river drainage, or a flock of hills, is more rare than a poet," he exclaimed. When he eulogized Archibald Marvine, he epitomized himself as well: "And so, fired with an enthusiasm for the discovery of the secrets of the mountains, he plunged into the wilderness far away from civilization."[24]

Powell conceived some genuine insights amid all this, which he fortunately prevented from evaporating into Hegelian mists. His geological contributions resided in a few dramatic insights which he invariably spun into complex systems of genetic classifications, usually labeled with Latin or Greek terms as though they were new species of foraminifers. His technique varied little from that of those who discovered and mapped sandstones or collected ammonites. His real accomplishment lay in dramatizing the role of fluvial erosion, already an American tradition, and in showing the value of landforms as a legitimate province for geologic study—in effect, discovering new phyla of phenomena for evolutionary geology. Like the river he rode to fame, Powell's geologic thought pivoted on the leveling power of erosion. The concept of baselevel was fundamental to his understanding of landforms and hence of geologic history.

It is not difficult to trace Powell's definition of baselevel to his fascination with maps, just as a few decades later the same preoccupation by William Morris Davis led to peneplains. Powell's real preoccupation, however, was with understanding that surface which the map recorded. For one thing, he conceived of the earth's crust as very thin, perhaps only 50,000 to 60,000 feet thick; hence, most geologic phenomena were essentially surficial. For another, the land surface was the interface between people and geology, as it was between biology and geology in general. Powell's concerns with land reform and his conversion to anthropology, as Goetzmann has pointed out, revolve about the problem of "adaptation—a good Darwinian word." Here again Powell subsumed human cultures and fossil records into the broader field of natural history and used the topographic map for synthesizing both.[25]

Powell extended these organizing techniques to his investigations of the American Indian: tribes were classified and mapped and placed on evolutionary chains. He thus stands easily as a contemporary of Lewis Henry Morgan and his friend, Lester Frank Ward. Like Ward (also a paleontologist, for whom Powell later found employment on the Geological Survey), Powell insisted that human evolution followed "psychic" factors rather than strict natural selection— we could control the direction of our evolution and the character of our adaptation to the natural world, a variety of neo-Lamarckianism similar to Newberry's. Powell's conservation reforms, like Ward's prototypic welfare state, were intended to insure this process. In both cases government scientific bureaus would oversee the work. In expostulating on his new fossil equivalents, Powell once exclaimed, "O that a pope would rise and a holy catholic church of geologists—a pope with will to issue a bull for the burning of all geological literature unsanctified by geological meaning." He never became pope, of course, only getting as far as the directorship of the U.S. Geological Survey, where he handled political reforms in somewhat the same vein as he had pleaded for scientific ones.[26]

It is no surprise that Powell's conservation program, as outlined in his 1878 *Report on the Lands of the Arid Region*, emphasized exactly the same factors as did his geological monographs: water and land. Both took meaning from an understanding of the land's surface and how its waters were distributed. The topographic map intersected both themes. It is no accident that when, as director of the U.S. Geological Survey, Powell tried to implement his reforms, the issue pivoted on the value of these maps. Given a topographic map, one could color in grazing and irrigation lands as one could the Silurian or the Triassic.

Powell gave to conservation what he gave to geology: simplicity and vision. There were others who produced maps of a similar type, Wheeler and Hayden among them. But Powell triumphed because he packaged his subjects into a form readily understood and publicized their urgency. Most of the conservation reformers of the Progressive Era radiated out from his circle. So successful were they at refashioning the landscape that those institutions which seem to stand as Powell's monumental legacy have become among the chief nemeses of modern conservationists. It is only one of the many ironies of his career that his program for reclamation has diminished the romance of the great river whose conquest lifted him to fame. That his was a marvelous achievement is undeniable. Yet he has been canonized with time and, like most saints, for miracles he never performed.[27]

Powell held Gilbert, as he attracted—or antagonized—most of his contemporaries, by force of personality. Although Powell was famous for his liberality in ideas—Gilbert wrote that he was "phenomenally fertile in ideas . . . absolutely free in their communication, with the result that many of his suggestions—a number which never can be known—were unconsciously appropriated by his associates and incorporated in their published results"—their flow was really a complex exchange. Determining the exact origin of particular concepts is probably both impossible and unnecessary. Before either man published on the plateau province, they were busy exchanging information and speculations—"swapping lies," as Gilbert phrased it.[28] Yet, in the interpretation of the Colorado Plateau, Gilbert is clearly the conduit between the Newberry (Ives) and Powell reports. Both he and Powell published their first books in the same year, and both have as their theme the structural peculiarities of the plateau. But, where both men commented on the differences distinguishing the Colorado Plateau from the Basin Range, Powell concluded with those differences, spinning a system of genetic classifications to differentiate between them, while Gilbert, scrutinizing their underlying mechanical cause, revealed a fundamental if unexpected similarity.

It is easy to neglect GK's contributions to this interchange, for he was indifferent to claims of priority in particular and self-effacing in general. And his 1875 report for Wheeler, of course, had nothing like the dramatic flair, suspense, and pulpit oratory that made Powell's *Exploration of the Colorado River* a popular classic. Yet Bailey Willis, recalling his years of service on the early national survey, referred to Gilbert as

> Powell's better half. Perhaps no one else ever thought of them in that way, but in constant relations with the two I learned to know how much Gilbert, the true scientist, contributed to the geological thinking of Powell, the man of action. I do not think that they themselves were conscious of the degree to which the latter absorbed and gave out as his own ideas that the former had silently passed through. But as Gilbert's assistant, I was sometimes jealous for that generous soul and devoted friend.[29]

In the final analysis, which man presented which specific facts or first insights hardly matters. Theirs was a powerful friendship that endured even beyond Powell's death in 1902, as Gilbert again assumed a role that began when he first arranged Powell's fossils. As executor of the will, he tidied up the estate, wrote and edited biographical memoirs, and attended to Powell's widow, Emma Dean Powell. In eulogizing Powell, he fervently concluded:

> To the nation he is known as an intrepid explorer, to a wide public as a conspicuous and cogent advocate of reform in the laws affecting the development of the arid West, to geologists as a pioneer in a new province of interpretation and the chief organizer of a great engine of research, to anthropologists as a leader in philosophic thought and the founder, in America, of the new regime.[30]

For neither man did their companionship alter fundamental convictions about the nature of geologic explanation. Though he praised them, Gilbert hardly used Powell's classifications, then descriptively, rarely genetically. They were not, he noted, "founded on principles of causation, and cannot therefore be assumed to be final." Similarly, because of his ties to engineering, Gilbert's conservation studies diverged from the reformist programs of his visionary friend. Nor did Powell borrow, if he even comprehended, Gilbert's methods. Where Powell, for example, conceived the Colorado canyons as the scene of heroic adventure and as a place where, "if strength and courage are sufficient for the task, by a year's toil a concept of sublimity can be obtained never again to be equaled on the hither side of Paradise," Gilbert discovered a different theme to present before the Philosophical Society of Washington. His address was titled: "On the Uses of the Canyons of the Colorado for Weighing the Earth."[31]

Although their social ties were less compelling, Gilbert's intellectual exchanges with Clarence Dutton were no less brisk. It was Gilbert who first escorted the survey's new recruit to the High Plateaus in 1875 and who passed on his own hard-won experience in field geology. With his training in physics and mathematics, Dutton could, in turn, converse with Gilbert on the larger questions of geophysics; with his expertise in chemistry, he could supplement GK's investigations into plateau volcanism with the solid petrology Gilbert lacked. In Dutton's military rank, earned through competitive examinations and some pretty rough service in the Civil War, the Major found a man after his own heart who would do for the plateau province what he had done for the canyon country. Together they established the literary genres for rim and river. A congenital polymath—he once called himself omnibiblical—Dutton's extraordinary versatility allowed him to create an aesthetic for the plateau based on a consideration of landforms as a species of natural architecture. In his first monograph, *Report on the Geology of the High Plateaus of Utah*, he exclaimed that on the Aquarius Plateau the geologist found himself a poet. Before he left the plateau, Dutton showed how exciting that amalgamation could be.[32]

Dutton's career was as unorthodox as his monographs. A Yale man, he was an athlete (he rowed crew), a litterateur (he won the literary prize as a senior), a military veteran who made the army his career, an accomplished raconteur, and an agnostic—after spending two weeks in a theology school where, in his words, he left before he was thrown out. His early interests centered on chemistry and explosions; his earliest scientific paper presented the first analysis of the Bessemer steel process given in the United States. Hence it was natural that, as an army officer, he should be interested in ordnance problems and, as a geologist, in volcanism. His memory was prodigious; most manuscripts were composed in his head, then written or dictated in marathon sittings. This procedure was, no doubt, one reason for the exceptional grace and conversational tone of his writings. Like Gilbert, he loved cerebral games, especially chess—although eventually he abandoned the game because the sleepless nights he spent puzzling over it threatened to undermine his health. Actually, as had Gilbert with mathematical riddles, he only transferred that passion to a new subject: geology.

In 1871 Dutton was reassigned to Washington, where he quickly insinuated himself into scientific clubs, especially the Philosophical Society, "cultivated the Survey men and became well acquainted with Powell." Thanks to Powell's connections with Grant, the War Department detailed Dutton out to the Powell Survey in 1875. The arrangement lasted fifteen years. Thus began a long furlough, a wonderful intellectual adventure for an inveterate cigar smoker who loved to read Macaulay and Twain and who, throughout it all—amid the most arduous fieldwork and the most sophisticated theorizing—maintained an irrepressible touch of irreverence. Not fully a professional by modern standards, yet far from being a mere dilettante, Dutton earned an international reputation through his writings which brought him election to the National Academy of Sciences in 1884. A man who valued cultured society, it was Dutton, moreover, who in 1878 suggested to Powell that they found the organization which became the Cosmos Club.

As his name for the club suggested, Dutton's was a romantic intelligence. What attracted him to a subject was a sense of its scope, vision, and unity. It was his desire, as he put it, "to discuss only such geological fields as present a series of facts which can be grouped together into a definite, easily comprehensible whole, and to avoid a subject which has, so to speak, neither head nor tail to it."[33] He found that "comprehensible whole" in evolutionary laws. Through them he could relate his favorite subject, volcanism, to igneous petrology, earthquakes, isostasy, and the larger questions of earth history. Through his analysis of volcanic processes, he was able to bring igneous stratigraphy and landform studies into conformity with theories on the geophysical evolution of the earth. As a result of expeditions to Hawaii, Central America, and the major volcanic fields of the Far West, he became America's outstanding volcanist, a theoretician in the field which underwrote the major laboratory discipline in American geology. At the same time, as a result of his famous investigation of the Charleston earthquake of 1886 and his 1904 textbook on earthquakes, he was credited with bringing the new seismology to the United States. He was equally renowned for his criticism of the contractional hypothesis and for his 1889 paper on the subject. In 1905 he rushed to apply the new crustal heat source discovered in radioactive decay to old problems in volcanism. For his labors Dutton became a major figure in the reconciliation of geological and geophysical theories of global evolution. Thus he belonged in the same tradition as James Dwight Dana, Joseph Le Conte, and Thomas Chrowder Chamberlin and, appropriately, it was George Becker, a leading American geophysicist, who memorialized him.

Yet it is probably for his landform studies on the Colorado Plateau that Dutton is best remembered. His achievement was two-fold. On the one hand, along with Powell and Gilbert, he brought the strange spires, majestic cliff façades, and fabulous canyons into the realm of scientific explanation; on the other, he gave them a critical aesthetic meaning. To overcome the linguistic poverty of English, he brought in new descriptive terms from Spanish and even native Hawaiian and scrapped stock Alpine analogies for striking allusions to Oriental architecture. The extraordinary land sculpture of the canyons demanded an elastic mind. That Dutton had, along with a prose style of great suppleness and infinitely varied rhythms. He differed from Powell in that he saw the morphology of scenery by analogy to art rather than to paleontology; he found an aesthetic meaning where Powell discovered a taxonomic order. He differed from Gilbert in that he measured the landscape with figures of speech rather than with numbers. But what really organized his influential studies of the plateau, as with his other investigations, was a commanding sense of history.

Dutton's study of the Colorado Plateau consisted of three monographs, which encompassed both its geologic and its scenic meaning. Each work had a common theme, erosion, and a common subject, volcanism. In the 1880 *High Plateaus of Utah*, volcanism dominated. In fact, a major cause for the elevation of the plateaus amid a landscape otherwise devastated by erosion was precisely their thick volcanic caps. Yet, in *Mount Taylor and the Zuñi Plateau*, officially published in 1886, Dutton discovered an inverse relationship: in a region formerly inundated by volcanic extrusions, erosion had progressed so far that only the vents and conduits remained, like broken ruins in bizarre testimony to past eons. In both studies, moreover, Dutton presented his material in the form of a journey—not a strict itinerary but an imaginative recasting of events, so that in the course of the journey one traveled through the geologic history of the region.

Between these two studies lay his masterpiece, the *Tertiary History of the Grand Canyon District*. Here there was a balance between volcanism (at Mount Trumbull) and erosion. Erosion was made a creative agency, an architect, and Dutton became its art critic. What appealed powerfully to him was the wonderful sense of gigantic proportion displayed by the region, the almost baroque splendor carved out of its grand symmetries. He called the Grand Canyon "a great innovation in our modern ideas of scenery" and, through his descriptions, assured himself a position as the great innovator in our interpretation of it. In two chapters, written from the vantage point of a promontory he named Sublime, he summarized the scenic evolution of a day and the physical evolution of the landscape. Thus he gave evolution an aesthetic dimension just as Newberry, out of a similar sense of earth poetry, had given it a theological one.

It is appropriate that the *Tertiary History* was published in 1881 as part of the U.S. Geological Survey annual reports and in 1882 as the first of the Survey's distinguished line of monographs. While mining engineer Samuel Emmons might grumble that he started Dutton's book "but came to the conclusion that life was too short," the monograph was well received.[34] It was received, in fact, much like Osmond Fisher's first synthesis of geophysical thought, *Physics of the Earth's Crust*, which was published contemporaneously with it, which quoted Dutton frequently, and which Dutton reviewed in the *American Journal of Science*. Fisher's text was an attempt to relate geophysical topics to the evolution of the globe made possible by the contractional hypothesis. Thus his interests and Dutton's were complementary.

It is appropriate, too, that the *Tertiary History* was published the same year that Herbert Spencer made his triumphal tour of the United States. Simply substitute "fluvial erosion," in the one case, and "contraction," in the other, to discover that Dutton and Fisher approximated the famous Spencerian summary of evolution as a progression from incoherent homogeneity to increasingly coherent heterogeneity. Dutton's narrative had advanced by stages to the grand climax at Point Sublime. In a century renowned for its great historians, he joined their ranks with his heroic biography of the Grand Canyon. Both he and Powell had taken the Colorado River as a guiding theme and shaped it into a scientific history—which was, in a sense, what geology (itself an alloy of discovery and the concept of evolution) did throughout the century. Thanks to their labors, in the Grand Canyon evolutionary geology found an unsurpassed summary of its lessons, and not since Dutton has the canyon found a more eloquent interpreter.

The Powell Survey never founded a formal school of thought, although its impact on the institutional composition and theoretical predilections of the U.S. Geological Survey was considerable. The academic synthesis of its work was left to the Harvard schoolmaster, Davis. Instead, its function was to dramatize new geographic discoveries and to establish conceptual links between its new theories and existing intellectual traditions.

Each member sketched those links in different ways. Indeed, their interpretations of the stratigraphy, geomorphology, and structural geology of the Colorado Plateau differed in much the same ways. What differentiated Dutton and Powell, in particular, from Gilbert was their historical sense, the evolutionary vista. Powell saw in the infinitesimal flexing of the Uinta Mountains the majestic patience of geologic processes, and in the mountain's stratigraphic profile he saw a regular succession of events which he could project as a blueprint for the future. Dutton had a complementary vision. In the vast layers of strata peeled from the surface by erosion and in the record of ancient erosion surfaces of similar magnitude, such as were manifest by the great unconformity in the gorge of the Grand Canyon, he saw the burden of the past. When he peered eastward from the Aquarius Plateau, he saw a landscape of sublime desolation, the eroded buttes and crags like fallen idols, the ruins of ancient geologic empires. To give them a head and a tail, he told their life history.

At the Henry Mountains, however, Gilbert interpreted the burden of history in a different sense. He measured the stratigraphy simply in terms of its dead weight, one of a variety of physical variables which entered the equation of laccolith formation, and he ignored the continuity of history. Instead, he exhibited the initial condition of the range as opposed to its present condition in stark juxtaposition, content to show the two periods as discontinuous systems, each obeying its own equilibrium. In a sense the laccoliths were timeless. The forces of past and future were treated like the weighted bob at the end of a swinging pendulum.

"Certain allied problems in mechanics": The Henry Mountains

Gilbert wrote four great monographs in his lifetime: *Report on the Geology of the Henry Mountains*, 1877; *Lake Bonneville*, 1890; *The Transportation of Debris by Running Water*, 1914; and *Hydraulic-Mining Debris in the Sierra Nevada*, 1917. The *Henry Mountains* is the prototype. Like all the rest, it is concerned with questions in geomorphology; like *Lake Bonneville* and the posthumously published *Studies of Basin-Range Structure* of 1928, it involves problems in structural geology. Alone of the group, the *Henry Mountains* addresses volcanism. In fact it virtually concludes Gilbert's thoughts on the subject, except for a few cursory references in *Lake Bonneville*, his explanation for Meteor Crater, and his study of the origin of the mcon. Yet, in the literature of American geologic exploration, there is nothing else like it: it stands as isolated and majestic as the Henry Mountains themselves, rising from the surrounding plain.

Of all GK's work, the *Henry Mountains* is the one most often celebrated and contested; it best illuminates those traits which made him at once representative and unique. It differentiated not merely between a choice of solutions but between a choice of techniques. The work has become symbolic, and the debate over it promises to continue; like mathematicians attacking Fermat's last theorem, each generation of geologists seems destined to break a lance or two over it. The *Henry Mountains* established a stylistic and conceptual pattern for the rest of Gilbert's major publications. At the same time, the record of its composition, the accounts of Gilbert's actual fieldwork, and the sources for the many threads of information and influence that he knotted into his explanation of the Henrys—all are better preserved than for any of his later masterpieces. The *Henry Mountains* presents some very unusual insights into the earth's mechanics—and into the mind of one of the earth sciences' great mechanists. It merits a close textual and conceptual analysis.

Perhaps the most salient fact about the *Henry Mountains* is the speed with which Gilbert produced it. He examined the peaks in the field for a little more than one week in 1875 and for a little more than two months in 1876. Most of those weeks were spent in topographic mapping. He prepared the report in three months' time, and much of that went into constructing plaster models and stereograms. Only a lean 150 pages long, the monograph—clearly a tour de force—was highly concentrated in its ideas and crystalline in its composition. The conclusion is inescapable that Gilbert had a sharp conception of the process of mountain formation and its erosional dynamics at the Henrys from very early in his research, if not before.

As he conscientiously noted in his preface, Gilbert was not the first to see the Henry Mountains or others of related type. During his 1869 voyage, Powell had named the peaks after physicist Joseph Henry, the first secretary and director of the Smithsonian. A. H. Thompson had blazed a trail to the region from Kanab in 1872; E. E. Howell had passed close by in 1873; and Gilbert himself had probably viewed them from the Aquarius Plateau in 1872. Meanwhile the Hayden Survey, operating east of the Colorado River, had begun publishing reports in which the Sierra Abajo, El Late, La Sal, and other mountains genetically similar to the Henrys were described. William Holmes, a member of the Hayden Survey whom Gilbert knew well by 1875, had written an analysis of the La Plata Mountains using the idea of an igneous intrusion which deformed country rocks; and Archibald Marvine, Gilbert's talented brother-in-law, had also commented on these mountain forms so peculiar to the Colorado Plateau. And, prior to anyone else, the ubiquitous John Strong Newberry visited them. His Macomb report did not see publication until 1876, but he certainly communicated his ideas on this, as on other matters, to his protégé over the intervening years. It is obvious that the laccolith as a mountain form, and perhaps as a mountain-building process, was on the verge of independent discovery by half a dozen geologists, and after Gilbert published in 1877 a dispute over priority predictably developed. GK's own experiences on the Wheeler Survey, furthermore, prepared him well: the *Henry Mountains* opens with a comparison between a volcano and a laccolith, a contrast he developed in his major report for Wheeler by pairing the Zuni Plateau with Mount Taylor. What is less obvious is the peculiar analysis he gave his discovery: by announcing a new form of analysis more than simply a new form of mountain, he inaugurated another, more fundamental dispute which still continues.

In structure, Gilbert's monograph is simple, consisting of three nearly equal parts. The first part systematically describes the geography of the Henrys; this was the first methodical survey of the region, the one which brought it into the realm of formal knowledge. The second part explains the formation of the Henrys, a range which represented a new form of mountain structure. Gilbert named this type of mountain the "laccolite," meaning "lake-rock." (To avoid confusion with minerals, Dana altered the Greek to "laccolith," though Gilbert persevered in his original spelling.) The laccolith thus became the second of his newly proposed mountain structures, the other being that of the Basin Range. In his technique for analyzing the mechanics of laccolith formation, moreover, Gilbert furnished a fundamental example for the solution of geological riddles by analogy to mechanics and mathematics. The third part of the monograph, explaining the existing processes of land sculpture acting on the Henrys, systematically codified the mechanics of erosion and added a basic contribution to the American fluvial tradition by describing the dynamics of what later became known as the graded stream. Thus Gilbert gave a dynamic explanation for the new mountains and landforms of the West where his survey director, Powell, gave a taxonomic order and his associate, Dutton, a historic context. After he finished, his reputation—and that of the Powell Survey—was assured. As Charles Hunt has remarked, "For more than 60 [now 100] years the Henry Mountains have been referred to in the geological literature of every language and are one of the localities most widely known to the science. No geologist needs to be introduced to them."[35]

Gilbert's analysis of laccolith formation developed from a key observation, a pair of assumptions in rheology, two mechanical analogies, and a fundamental comparison to volcanoes. The contrast between the volcanism of the Henrys and that typical of most volcanic piles initiated the presentation. What made the *Henry Mountains* more than a mere recapitulation of Gilbert's report for Wheeler was the critical observation that there existed a correlation between the size of a laccolith and its depth, with the corollary observation that there were no small laccoliths. The rising column of magma thus behaved according to the principle of hydrostatic equilibrium: it ascended through the rock until the point at which its driving pressure equaled the weight of the overburden was reached; then it began to spread laterally as a sill, in an approximately circular pattern. Gilbert imagined the sill as a hydraulic press, trying to raise its overburden. This required that it have a certain area and that this necessary area increase with the depth at which the sill was buried. The deeper the sill, the larger the limital area. When the sill became sufficiently large, a laccolith resulted.

To calculate the forces involved, GK converted the hemispheric shape of the laccolith into a cylinder. This he did through a clever pair of assumptions. He likened the laccolith dome to a monocline—a simple steplike fold—rotated about a point. The monocline, in turn, was structurally homologous to a fault—a crustal fracture with one side of the break moving relative to the other. Hence, one could replace the circular monocline with a circular fault, and the laccolith became a mountainous piston. Gilbert concluded his analysis, however, by returning to the case of the dome. "Guided by certain allied problems in mechanics"—he cited Rankine's *Manual of Applied Mechanics*—he likened the stresses in the flexed strata to those involved in the bending of a structural beam. Considered in thermodynamic terms, the resulting laccolith was analogous to a volcano formed under high confining pressure. Precisely because this pressure was so intense, the strata behaved plastically rather than brittlely under the stress of flexure. Otherwise a series of fractures and dikes would have spilled the magma onto the surface, creating a more familiar volcanic landscape of cones, piles, and lava flows.

Gilbert summarized the shaping of the laccolith in this way:

The laccolite in its formation is constantly solving a problem
of "least force," and its form is the result. Below, above, and on
all sides its expansion is resisted, and where resistance is
greatest its contour is least convex. The floor of its chamber is
unyielding, and the bottom of the laccolite is flat. The roof
and walls alike yield reluctantly to the pressure, but the
weight of the lava diminishes its pressure on the roof. Hence
the top of the laccolite is broadly convex, and its edges acutely.
Local accidents excepted, the walls oppose an equal resistance
on every side; and the base of the laccolite is rendered
circular.[36]

What began as a systematic contrast to the mechanics of a surface
volcano ended as an original insight into a new form of mountain
structure.

This structural analysis brought a novel perspective to the rugged mountains of the West. Where others, such as Powell and Dutton, had brought geologic views hybridized with aesthetics, literature, and biology, Gilbert brought a perception derived from physics
and engineering. He also showed the advantages of his mathematical treatment. On the one hand, his mathematical-physical model
furnished a concise logical set of relationships among the data; on
the other, it allowed for predictions which were not inherent or obvious in those data. In concluding his description of the laccolith,
Gilbert offered exactly such a prediction. Trying to estimate the
depth of overburden (all of it Tertiary strata) which had been removed by erosion since the formation of the laccoliths, he arrived at
a figure of 7,700 feet. The enormous scale of erosion whose discovery by earlier explorers to the Colorado Plateau, such as Newberry,
had startled the scientific community, the ruinous decay of landscape on which Dutton and Powell had lavished their purple prose,
now had a quantitative meaning.

The final and third part of the *Henry Mountains*, perhaps more
than any other work of the century, erected the foundations for
modern geomorphology. Under the general rubric of land sculpture,
Gilbert described the erosional dynamics operating on the modern
Henrys. He divided this analysis into three sections, each bringing
the subject to a more elaborate stage. The procedure demonstrated
the same logical style he had used in building his model of laccolith
development. First, he codified the erosive processes and the manner in which environmental conditions, such as slope, lithology, and
climate, modified their rates. Second, he tried to axiomatize these

processes, especially fluvial erosion, into a set of three laws which governed land sculpture. Third, he integrated land sculpture, again through rivers, on the regional scale of the drainage basin. In each case examples were drawn from, and conclusions applied to, the Henrys.

Gilbert modestly claimed that he was only restating certain "principles of erosion which have been derived or enforced by the study of the Colorado Plateau."[37] In one respect this was true. Much work had been done on rivers, though the bulk of European geologists considered stream erosion to be a trivial geologic phenomenon; consequently, most fluvial studies belonged with physics in the form of hydraulics and hydrodynamics or with engineering in the investigation of canals and irrigation systems. The pioneer engineering work was done by the French and Italians in the eighteenth century, while the canon of physical studies would come together in 1879 with the publication of Horace Lamb's *Hydrodynamics*. But American geology had a distinguished tradition of fluvialists— Dana, carving out a reputation with South Sea islands; James Hall, with Niagara; Andrew Humphreys and Henry Abbot, with the Mississippi; Newberry and Powell, with the Colorado. It was not the novelty of his particular ideas that made GK's own report remarkable but its brilliant codification of known processes and the axiomatic format in which they were presented. Actual observations were verified by derivation from these axioms as though they were lemmas to a geometric theorem.

Of Gilbert's three laws of land sculpture, the first was the law of declivities, which derived from the evident fact that "if steep slopes are worn more rapidly than gentle, the tendency is to abolish all differences of slope and produce uniformity."[38] The second law was the law of structure, derived from the differential erodibility of rocks. The law of divides, the third law, held simply that the grade of a river steepened as the divide was approached.

In themselves the laws are simple. But it was Gilbert's special insight to set them into mutual opposition: "The law of declivities thus opposes diversity of topography, and if not complemented by other laws, would reduce all drainage basins to plains." However, this required a uniformity of conditions which nowhere exists, so the law is opposed by the law of structure, which generates variety of form. This action, or differentiation, "continues until an equilibrium is reached . . . When the ratio of erosive action as dependent on declivities becomes equal to the ratio of resistances and dependence on rock character, there is equality of action." Similarly for the law of divides. Thus, the laws of declivities and divides are responsible for the erosive forces, the law of structure for the resistant forces, and the actual landscape represents a dynamic equilibrium between their antagonistic tendencies. The formulation is virtually identical to that used by Gilbert in describing laccoliths and streams.[39]

Just as his analysis of the laccolith began with the comparison between Mount Taylor and the Zuni Plateau, so did Gilbert's analysis of stream action begin with another contrast drawn from comparative geography. In this case, he took the different phases of streams commonly recognized in natural history: the torrent phase, characterized by erosion; the river proper, typically engaged in the transportation of water and debris; and the delta, characterized by deposition. What Gilbert achieved in the *Henry Mountains* was to bring both to the structural problem of laccolith formation and to the geomorphic question of stream behavior a similar perception of nature as well as similar techniques of analysis. From geographic features he created a dynamic model based on physical laws and organized around the concept of equilibrium. For the laccolith, this involved hydrostatic equilibrium; for the stream, hydrodynamic equilibrium. In framing his data with conditions of equilibrium rather than evolution, he extended his analysis beyond the range of natural history and intellectually divorced himself from his contemporaries. Where Dutton narrated an evolutionary structural history of Mount Taylor and the Zuni Plateau, Gilbert reconstructed the physical conditions which distinguished the subsequent forms adopted by the rising magma. Where Davis took the three phases of rivers and made them into sequential stages of river evolution, a transformation over time, Gilbert redesigned them into an "equilibrium of action" which tended to be uniform over time. In both cases, he redefined the meaning of "geologic system"; rather than a historic stage, it designated a dynamic unit of geologic processes in equilibrium. Where Dutton and Davis, for example, converted geographic differences into historic systems, separate features representing chains in an evolutionary progression, Gilbert fashioned them into components of a dynamic system, operating in a steady state.

Gilbert conceived the stream as a system of energy, as an engine which performed work according to the laws of thermodynamics. Although his discussion was qualitative, it had a highly structured, almost Euclidean logic guiding it; and at its base were two considerations: the law of the conservation of energy and the principle of least action. According to Gilbert, the stream existed in a state of dynamic equilibrium. It adjusted its longitudinal profile, or grade, so that its energy was sufficient for the work demanded of it. Put differently, his concept held that the energy of the stream tended to be uniform throughout its length or that the stream channel represented a balance between force and resistance. As the amount of debris or water to be moved or as the resistance to movement varied, so did the stream's grade; thus the stream had just the velocity it needed to transport its materials. The result, although Gilbert did not use the term, was a graded stream.[40]

It is obvious that Gilbert sought to organize his explanation of the Henry Mountains by analogy to mechanics. What is less obvious is his indifference to the great mechanical law that proved so important to the geological thinking, especially the geophysical thinking, of the latter nineteenth century: the second law of thermodynamics. He organized his description of the Henrys, both as they were formed and as they exist today, by presenting them as separate systems in equilibrium. To his contemporaries, the more important facts would have been the sequences of rocks piled up or swept away, the degradation of energy involved in the orogenic events, the systematic denudation of the landscape which would have followed, the placement of laccolith creation within the great evolutionary chronology of the earth—in short, the life history of the mountain and of the rivers which drained it. Gilbert not only ignored the degradation of the systems over time but, in his enthusiasm for his ideal model of laccolith formation, nearly forgot to mention just when the event occurred.

In this way, GK brilliantly dramatized two concepts which underpinned his interpretation of nature: equilibrium and rhythmic time. The frontispiece to the *Henry Mountains*, for example, shows two historical systems—each complete in itself, without antecedents, consequents, or transitions—juxtaposed into the same block diagram (incidentally, though he did not invent the block diagram, Gilbert gave it new possibilities with this drawing that were soon exploited by others in the West). In his text Gilbert explained how the two sets of forces, one in each system, acted against an opposing resistance to create the form shown in the diagram. The magma column and the stream had each, in its time, reached a profile of equilibrium in acting against the resistance offered by the rocks of the region. In both cases a path of least force was taken. In the case of laccoliths, the resulting equilibrium was static; in the case of streams and landforms, it was dynamic. As long as there existed an equilibrium of action, the ratio of force to resistance effectively shaping the landscape remained constant. As long as this ratio remained constant, there was a similar conservation of form, whether

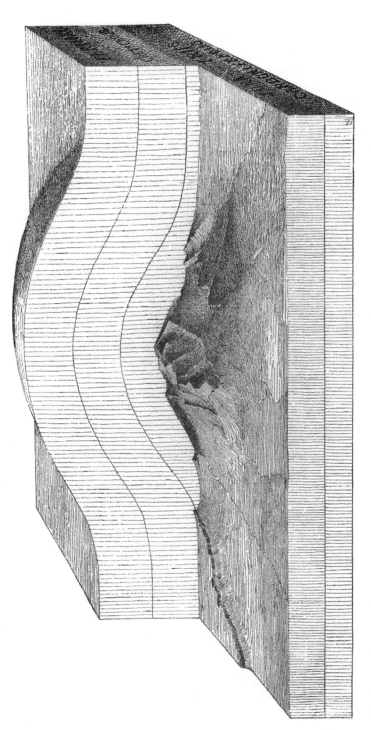

The frontispiece to the *Henry Mountains*—"half-stereogram of Mount Ellsworth [in the Henrys], drawn to illustrate the form of the displacement and the progress of the erosion. The base of the figure represents the sea-level. The remote half shows the result of uplift alone; the near half, the result of uplift and erosion or the actual condition."

as the profile of a river, the dome of a laccolith, or the face of a cliff. When Gilbert depicted some of the streams draining the slopes of the Henrys as shifting their channels laterally, the river—like the laccolith—was only solving the problem of least force.

GK's description of lateral planation by streams was probably the first in geologic literature, but it followed almost as a corollary from the peculiar physical conditions of the Henrys in which the streams worked, just as the laccolith rather than a volcano followed from the circumstances of magmatic intrusion. Lateral planation was not a stage in the evolution of a river any more than the laccolith was a stage in the evolution of a volcano; the two systems are nowhere joined by a historical narrative describing what happened, era by era, between the time the laccoliths were formed and the present. Gilbert's philosophy of history thus shaped his monograph, allowing him to ignore precisely those questions typically asked most eagerly by geologists of his day, namely, the age of the events and their systematic change over time.

Equally instructive about the *Henry Mountains* are those things Gilbert did not do. There is an almost total indifference to historical geology and to contemporary interpretations of volcanism. His insights were physical rather than chemical, dynamical rather than topographical, typological rather than historical. Especially revealing was his neglect to describe the historical erosion of the thousands of feet of strata that, to the south, awed Dutton, Newberry, and Powell—the very phenomenon that was, in fact, at the heart of their monographs on the Colorado Plateau. Gilbert's geologic systems were cross sections through time, not longitudinal profiles. Absent, too, was a certain Humboldtean fever one sees in Powell's excitement over thousands of feet of unnamed strata or in Dutton's restless eye, scanning with the acumen of a connoisseur the proliferation of unnamable shapes that decorated the desolation. Gilbert's impulse was to find the essence of a scene, not to celebrate its variety. And, finally, there is evident again that engineering imagination which sought a solution for the question at hand but no more. There is no effort to universalize from the laccolith into a general theory of volcanics or of orogeny; an ideal model of laccolith formation was sufficient. To expect further generalization would be like asking the boy who solved the riddle of the bread loaf to create from its solution a general mathematics of hemispheres. The problem did not call for it.

The danger in Gilbert's work was never that he would generalize too broadly beyond the problem at hand but that he would fail to piece the individual puzzle parts into a larger mosaic, that he would merely solve a number of particular riddles in the vogue of an engineer. The *Henry Mountains*, like all his other great works, was spared both extremes. Guided by its theme of equilibrium, it seemed itself to show between the driving force of its generalizations and the opposing resistance of its data a dynamic equilibrium of insight.

The *Henry Mountains* was one of the most suggestive geologic works of the century, and, like most important studies, the response to it varied widely. Again, most Europeans—particularly Germans—dusted off their national authorities, as scholastics might have produced Aristotle, to demonstrate its fallacies. They protested that the volcanics had formed first, with sediment later deposited around them and taking on the shape of a dome. This was an argument Gilbert had anticipated and devastated in his text, but, as with his announcement that vertical forces elevated mountains, it took several decades for general acceptance in Europe, despite help from a few friendly critics like Archibald Geikie.[41]

Much the same occurred in America. Some saw in the report new categories of geologic phenomena to search out, identify, and map. Soon "laccoliths" and "pediments" (based on Gilbert's description of terraces formed by lateral planation) dotted geologic maps. The fact that these forms were particular cases of general laws, mere beams in an elaborate edifice of mechanical analogies, escaped many, who read the report as announcing a discovery—a new type of mountain form or, at most, a new mountain-forming process—not as a new way of looking at mountains. In short, the work was usually read as a high-class exploration account which had to be absorbed into the reigning paradigms, especially classification schemes and evolutionary histories. Hence, the laccolith took its place among Powell's genetic taxonomy of mountain forms and the graded stream became a moment in time in the evolutionary history of a river.

There were problems of priority and personality, too. James Dwight Dana exercised profound skepticism, for the laccolith did not conform to his theories of volcanics. Consequently, when laccoliths finally appeared in his influential textbook, they were explained solely by virtue of viscosity effects. The disagreement is not surprising: Dana came from a background in mineralogy, Gilbert from mechanics. Meanwhile, the Hayden Survey, as self-aggrandizing as ever, rushed an article into print asserting priority to the discovery. GK countered by inserting a last-minute footnote to his text, merely summarizing the claims without comment. He needed to say nothing more, for the mechanisms were sharply different. The Hayden geologists thought the intrusion was primarily thermal, melting its way up through the rocks and incidentally bulging them, while Gilbert thought the process was determined primarily by pressure, pushing up with little thermal exchange. Also, the Hayden geologists envisioned the intrusives as gradually working their way upward to the surface, while Gilbert, though he found dikes radiating out of the domes, designed a model which required a high confining pressure to prevent the strata from behaving in brittle fashion or allowing the dikes to rise to the surface.[42]

But to the end of his life William Holmes, who went on to a distinguished career as artist, anthropologist, and administrator, felt that he had been swindled out of his personal discovery. He did not really blame Gilbert as much as he did later historians, but he was only partly appeased by Gilbert's gracious tribute to him in the *Henry Mountains*. Gilbert had named two of the laccoliths after him, a Greater and a Lesser Holmes. But, since the area was still being excavated by erosion, GK wrote with unintended irony: "In attaching to the least of the peaks the name of my friend Mr. Holmes, I am confident that I commemorate his attainments by a monument which will be more conspicuous to future generations and races than it is to the present."[43]

Gilbert's allusion to "certain allied problems in mechanics" was not a casual, parenthetical phrase. In one respect, it symbolically summarized the progress of the earth sciences. Whereas James Hutton, experimenting with the analogy of the earth to a heat engine, could imagine volcanic features only as safety valves or ruptures, geologic events which did not perform mechanical work, Gilbert, drawing on the growth of physics, could picture the laccolith as a piston, an integral part of the earth's machinery, operating according to regular laws and with measurable efficiency. In another respect, the phrase epitomizes the unique position he occupied in American geology. At one point in his report for Wheeler, he had posed a question in structural geology and remarked: "Of late years the most important contributions have come from the physicists, and in their scales have been weighed the old theories of geologists. Here will be an opportunity to compare the speculations of the physicists with new geological data."[44] In a large sense that was exactly the task he performed so superbly at the Henry Mountains and exactly why there are many today who say that twentieth-century

American geology, both structural and geomorphic, began in 1877 with a monograph whose outlines were confidently sketched during a freezing October rain in a tent pitched on the slopes of Mount Ellen.

Languages for a New Geology: Mechanics, Mathematics, Literature

The sense of equilibrium so strong in Gilbert's characterization of laccolith formation and the behavior of the graded river has equivalents in practically all of his other writings. At Lake Bonneville, one finds the graded beach; in the Sierras, the graded hillslope; and outside the Golden Gate, the graded tidal bar. This perception of equilibrium was a philosophical conviction. Where he could not name specific forces, he nevertheless held that the landscape was a product of two competing tendencies: one created diversity, the other uniformity. This was a vision that had more in common with Newton's depiction of planets, whose orbits inscribed a compromise pattern between two forces, than with Darwin's struggle for existence, in which, in a sense, one force overcame another. In fact, Gilbert even conceived of organic evolution in terms identical to those he used to describe landforms: "Thus the secular evolution of species," he wrote, "combined with the secular and kaleidoscopic revolution of land areas, leads to two antagonistic tendencies, one towards diversity of life on different parts of the globe, the other towards its uniformity."[45] The apparent progressiveness evident in the fossil record was an illusion. So, too, was its geophysical counterpart: the inevitable, systematic decay of physical landscape.

This perception of equilibrium gave Gilbert an organizing concept. Nearly all his monographs, for instance, begin with a disequilibrating force and end when equilibrium has been reestablished in the region. This same appreciation guided his choice of mechanical analogies, his preference for certain kinds of illustrations, and even his style of mathematical and literary expression. GK was a pure Newtonian who sought to explain geologic phenomena by bold mechanical analogies, and this orientation gave him a curiously conservative flavor. His impulse was to describe new marvels by analogy to older systems of knowledge, not, like Powell or Davis, to invent new sciences in response to a growing information base. And a similar classicism is manifest in his other modes of expression.

This mixture of new and old suggests an interesting parallel between Gilbert's theoretical investigations at the Henry Mountains and those of another titan of American science, physicist Josiah

Willard Gibbs. Both men published their first major studies in the 1870s, with Gibbs' monumental *On the Equilibrium of Heterogeneous Substances* being printed in nearly the same year as the *Henry Mountains*. The achievement of both men was to introduce thermodynamic concepts into new fields—Gibbs into physical chemistry and Gilbert into physical geology. The result was to give a dynamic basis to disciplines faced with an analogous conceptual and methodological problem—the preoccupation by chemistry with filling out the periodic chart was similar to the concern by geology with completing the stratigraphic column. Both Gibbs and Gilbert offered an alternative foundation to their sciences by organizing their data around the concept of an equilibrium system, and the outcome of adjustments in the state variables defined, for both the chemical and the geological systems, a surface of equilibrium. Similarly, the fate of both the phase rule and the graded stream concepts was to see the processes for describing states of matter turned into stages of evolutionary history. Thus, Davis' conversion of Gilbert's landscape studies had its parallel in Norman L. Bowen's use of the phase rule to describe the evolution of igneous rocks and in Henry Adams' famous essay, "The Phase Rule Applied to History."

The comparison extends even to the temperaments of the two men. Yet in the end the differences between them are perhaps more significant in placing Gilbert in the spectrum of nineteenth-century physics. For one thing, he never experienced the relative obscurity of Gibbs. For another, even when using physics Gilbert approached his subjects from the viewpoint of a geologist, while Gibbs, even though working in fields like chemistry, always addressed them like a physicist. Consequently, Gilbert never did anything in mathematics or physics comparable to Gibbs' rigorous advances in clarifying and elaborating fundamental physical laws. Gilbert never invented new mathematical procedures, such as vector analysis, and he never revised thermodynamics at its foundations, such as Gibbs did with his *Elementary Principles in Statistical Mechanics*. Instead he looked upon physics as a given body of techniques and concepts by use of which he could solve geologic problems. Most of what he read in physics, moreover, involved mechanics, and that had a peculiar bias. Whereas Gibbs looked for inspiration to the work of Rudolf Clausius, William Hamilton, Hermann von Helmholtz, James Clerk Maxwell, and Ludwig Boltzmann, Gilbert turned to William Rankine and perhaps Kelvin. His conduit for information on physics was the Scottish school, not the German; his approach allied physics more to engineering than to philosophy.

The influence of Rankine, in particular, may be taken as symbolic. GK never spelled out his larger intellectual mentors, except his admiration for Lyell, but, from his repeated allusions to the members of the Scottish school and from references to Rankine's celebrated *Manual of Applied Mechanics,* one can trace out suggestive associations. In Rankine's manual, Gilbert discovered a useful compendium of concepts and techniques, a methodology, and a philosophy of science—it was from the manual that he drew the analogy to "certain allied problems in mechanics" which allowed him to fashion his model for the laccolith. The manual long continued to serve as a reference book for that branch of physics which most attracted him, mechanics, as modernized by developments in classical thermodynamics. Gilbert remained oblivious to theories which dealt with more intangible or mathematical fields, such as electromagnetism, or with post-Newtonian developments, such as atomic physics.[46]

In the introductory essay to the manual, titled "The Harmony of Theory and Practice in Mechanics," Rankine gracefully critiqued what he termed "the fallacy of a *double system of natural laws"*—

> one theoretical, geometrical, rational, discoverable by contemplation, applicable to celestial, aetherial, indestructible bodies, and being an object of the noble and liberal arts; the other practical, mechanical, empirical, discoverable by experience, applicable to terrestrial, gross, destructible bodies, and being an object of what were once called the vulgar and sordid arts.

He was lamenting the schism between science and engineering, but to Gilbert it must have seemed that Rankine's double system equally described the division between natural philosophy and natural history. Rankine saw his popular manual as a mechanism for unifying engineering and science by putting both on a set of similar assumptions and techniques—thus it is entirely apropos that Gilbert should draw on analogies taken from the manual to underwrite his own hybridization of mechanics and geology. Where Rankine sought to describe mechanical and heat engines in terms of the new energetics, Gilbert applied almost identical techniques to the "physiographic engines" of the earth. Hence, it is no surprise to learn that the categories of scientific knowledge described by the two men differed only in terminology. Rankine resolved "mechanical knowledge" into three branches: "purely scientific knowledge—purely practical knowledge—and that intermediate knowledge which relates to the application of scientific principles to practical purposes, and which arises from understanding of theory and practice."[47] Gilbert said the same thing, distinguishing between a "pure," an "applied," and a "practical" science.

The manual also suggests the general nature of Gilbert's modes of expression. Take, for example, mathematics. Like his physics, this was closely tied to actual physical problems. The laccolith model is a case in point: it was fundamentally mechanical rather than mathematical, just as it was a geological problem rather than a purely physical one. For the most part, Gilbert's mathematics was that required by the manual—algebra, geometry, and some calculus. To this he added a few studies involving elementary probability. In all he found it easy to agree with Rankine's advice that, "as far as possible, mathematical intricacy ought to be avoided." In this way the concepts of the science would not become inaccessible to most of its practitioners. GK's fundamentally utilitarian mathematics easily conformed to this advice. Whenever he used a mathematical formula, he always tried to package it in terms meaningful to field geologists, as, for example, with his hypsometric tables and his isostatic unit, "rock-feet." His mathematics was simply part of the basic mechanical analogy he established to explain geologic phenomena. The fundamental perception was of a physical riddle which brought with it prescribed mathematical techniques rather than a mathematical relationship which, by analogy, could be applied to many physical situations.

Partly because he deliberately shunned technical virtuosity, it is uncertain how extensive Gilbert's mathematical knowledge was. Frequently, when stalled on some equation, he sought help from his friend Robert Simpson Woodward, a mathematical physicist, sometime employee of the U.S. Geological Survey and of the Coast and Geodetic Survey and eventually president of the Carnegie Institution of Washington. Woodward, for his part, expressed some amazement at the depth of Gilbert's comprehension of mechanics and mathematics while confessing that Gilbert never acquired much facility with the "higher mathematics." One reason for the deficiency was Gilbert's preference for interpreting abstract concepts into concrete, physical riddles, even when the concepts were mathematical. Thus, in determining the mean value for a scatter of points, he likened the problem to finding its center of gravity. Similarly, he once likened a set of curves to a topographic map. Combined with a sense of time that was equally concrete, it is little wonder that he imagined the derivative as an "instantaneous ratio." While this habit of mind made his thinking exceptionally clear and vivid, it also placed him outside the philosophical advancements that divorced mathematics from any necessary physical meaning and allowed Cauchy, earlier in the century, to rigorously define the foundations of the calculus. Gilbert's mathematics was thus exactly complementary to his physics. His choice of algebra over the calculus was equivalent to his interest in mechanics rather than in the electromagnetic field.

In one form or another, Gilbert's mathematics involved ratios. In his geological systems, the ratio functioned rather like an equilibrium constant in a chemical system. It was a broad measure of force and resistance, of uniformity and diversity. Thus, his mathematics resembled his scientific prose style. Corresponding to his prose digests of field observations, he could present a landscape of numbers—a mathematical stratigraphy of volumes and specific gravities rather than of fossils and mineralogical contents. So much was a concession to the accountant in him. Yet analogous to the literary high style he used in summarizing his thoughts was a synthetic mathematics which, again like his prose, worked best by generating comparisons. These relationships balanced as a ratio, however, rather than about a semicolon. Of course, not every mathematical expression consciously resolved itself into a ratio any more than every prose sentence affected a neoclassical construction or every conceptual explanation assumed an equilibrium condition. But these were the ideal, and Gilbert was most successful when they all meshed.

His use of ratios constituted another way in which he diverged from most of his contemporary earth scientists. Where, ideally, they looked to the calculus to describe geologic and geophysical processes, Gilbert preferred a more elementary algebraic ratio. Indeed, when he did use the calculus, he thought of it as a means to discover "instantaneous ratios," not as a measure of the rate of change between two variables. Since most mathematical investigations of earth processes analyzed them as they changed over time, and since the earth was imagined as a closed system, the calculus would theoretically be the ideal technique to discover the successive condition of the system at instants of time. Gilbert, however, rejected this concept of time. He thought of time as a ratio, intervals bounded by conditions of equilibrium, not as a continuous and progressive change. Since time was not an arrow, mathematical functions relating it to other variables were discontinuous and the calculus was not suitable. When he used differential equations, as in his flume experiments, time was not one of the independent variables. For those in evolutionary geology or geophysics, the analogy of the calculus to geologic time was a close one. The essence of the uniformitarian position, of course, was that tiny, infinitesimal forces acting over long periods of geologic time could sum up to significant changes in the landscape. The calculus was the mathematical equivalent to the geological and geophysical perception of infinitely small, continuous forces which could be isolated for some instant or summed up over eons of time. Thus, Gilbert's mathematics, like his conception of time, brought him closer to Archimedes than to Kelvin.

GK had other reasons to suspect the calculus. Conceiving of causality as a plexus of antecedents and consequents and insisting that all scientific observations were biased and imprecise, he had little use for the exactitude and apparent determinism inherent in the great millstones of the calculus. On the contrary, his ratios tended to be mathematical equivalents of such broad statements as, for example, the fact that there were lake-creating and lake-destroying processes. Besides, it is equally uncertain whether a mastery of more complicated mathematical technique would have assisted him much. Field geology was hardly a laboratory science, and his data were too crude to subject quantitative hypotheses based on them to tests which involved much less than an order of magnitude. The wide variance in the calculation of the earth's age, after all, did not offer an encouraging example.

Gilbert also issued a caveat in the *Henry Mountains* which would be well to quote from. "It is always hazardous," he warned, "to attempt the quantitative discussion of geological problems, for the reason that the conditions are apt to be complex and imperfectly known; and in this case an uncertainty attaches to the law of relation, as well as to the quantities to which it is applied."[48] Hardly a geological situation he investigated throughout his career offered any greater promise of perfection. The data were too gross to warrant too fine a technique; he would have been measuring miles with a ruler marked in inches.

If Gilbert's utilitarian orientation and his preference for tying mechanical models to particular geologic situations prevented him from joining the ranks of the sophisticated geophysicists in the British tradition, he was also spared their errors and presumptions. The result was an enviable flexibility, manifest as a talent for translating physical models, along with their accompanying mathematics, into geological situations. No explanation of a geologic event was complete until it assumed a mathematical-physical form; but no equation, no deduced quantity, could repeal a geologic fact. The man who had measured the Cohoes potholes and who scoured the Sequoia groves for a parabolic spider web was hardly hostile to mathematical analysis. But Gilbert, for all his training in mechanics, was not a physicist or a mathematician but a geologist.

Considering that his field of study was, after all, geology, Gilbert's mathematical and physical tools were quite satisfactory. Moreover, mathematics entered his work in other forms as well. For one thing, whenever possible he presented his observations in a deductive, almost Euclidean format. Such, indeed, was the heart of his "systematic approach." When he could not write quantitative relationships (as with the flume experiments he conducted in later life), he would imitate a mathematical structure in his sentences, almost to the point of writing prose equations. And, finally, there was almost an instinct in Gilbert to count. To match the engineer in his temperament there was an accountant—a congenital bookkeeper. Consequently, a sense of accounting—and of accountability—runs through his life. Instinctive though it seems, this habit harmonized with his conception of science, which, he insisted, differed from the arts by dealing in quantities, not qualities. This belief, in turn, supported his mechanical model for geologic processes. "Whenever a tentative theory involves the application of force or the expenditure of energy," he decided, "the investigator (or his critic) habitually asks whether the assumed cause affords a sufficient amount of force or of energy." He added that "quantitative tests of this particular type are among the most familiar resources of investigation." At one point, he even conceived of scientific culture as a bank. After ultimately identifying "mankind, collectively, through the agency of its men of science and inventors . . . as an investigator, slowly unravelling the complex of Nature and weaving from the disentangled threads the fabric of civilization," he phrased his conclusion in the imagery of accounting: "Knowledge of Nature is an account at bank, where each dividend is added to the principle and the interest is ever compounded; and hence it is that human progress, founded on natural knowledge, advances with ever increasing speed."[49]

On another, somewhat less mechanical level, GK's instinct to count, measure, and weigh and to discover fundamental geometric forms and mathematical relationships in nature elevated his use of mathematics beyond the status of a mere instrument or a reflex. That urge corresponded to a philosophical assumption that nature was, at heart and if approached in the proper way, simple and that, as Newton had, the skillful scientist could express the elementary laws which governed nature as a mathematical formula. This sense of mathematical form and logical order helped give stability and shape to Gilbert's mental equilibrium as fully as to the thematic structure of his monographs—it fashioned the axis of his psychological gyroscope. At its worst, though he never surrendered a basic compassion that spared him from dogmatism, that axis could misguide what he called his mental compass. Once in Philadelphia, for example, his train arrived at night at a station different from the one he supposed, and when he set off to navigate the town, having already programmed his route, he quickly became lost. In the world of people, whether the city or the office of a survey administrator, that compass could be similarly deflected by the field of his logic. But, operating in the world of nature, it rarely failed him.[50]

Even Gilbert's choice of illustrations tends to confirm an engineering bias. By replacing the topographic or geologic map with the block diagram or with plaster models and stereograms, he showed how he thought of landforms in terms of volumes and masses rather than simply sequences of surface shapes; he thought of geomorphology in the same terms he used for structural geology. These visual expressions were not far different from the kind that appeared in Rankine's manual. Nor did they demonstrate the historic evolution of environments and geologic forms which was so characteristic of the geologic thinking of the day and which—in the hands of William Morris Davis—would become a visual analogue of the calculus. Gilbert's drawings remained the sort that would appear in a text on applied mechanics.

But it was the classicist in Gilbert that organized his dynamic landscape without recourse to historicism or to its literary counterpart, the narrative. In place of historical stages, he substituted states of equilibrium and, in place of narrative prose, he constructed a neoclassical style that expressed the condition of equilibrium he perceived in the landscape. In his prose as much as in his conceptual organization of nature, no one theme or hypothesis or expression or force was allowed to overwhelm or subsume another; each topic was always regulated by an antagonist. In nature, this resulted in an equilibrium of action; in prose, in an equilibrium of composition. It is the style not of a man announcing discoveries but of one arranging systematic comparisons and marshaling syntheses. Like the explanations it transmits, this prose is rarely remarkable for its virtuosity or technicality, but it *is* remarkable for its clarity. Like his understanding of physical nature, Gilbert's prose shows a profound appreciation for balance and structure. His was a style perfectly machined to present systematic contrasts, just the tool needed by a man who saw nature as a plexus of competing processes.

Whether comparing two rivers or two hypotheses, Gilbert's approach was identical. His stylistic and conceptual formula called for balancing them equally, as though the sentences in a paragraph or the clauses in a sentence were placed on two sides of a fulcrum or, given his penchant for quantification, as though the words and thoughts were literally being weighed on a delicate scale. His arguments do not propel one inexorably forward to a predetermined conclusion but come as though he were simply reporting the relative weights from a scale. His careful constructions were shaped not for dramatic effect, like Powell's, or for the display of suppleness and variety, like Dutton's, but for the ability to balance and weigh. Judgments and evidence pivot about a clause or a semicolon as about the fulcrum of a scale. His conclusions come as dispassionately and inevitably as an arithmetic sum. There is a cadence to much of his less formal writing that ties his prose together as rhythms do the geologic events they portray.

GK was not a great literary stylist or an outstanding mathematician or an ingenious, if misplaced, physicist. He was a geologist of great breadth and insight. His skills with mechanical analogies and his techniques with mathematics, prose, and illustration he used to express his vision as a geologist. All share in his general conviction of equilibrium in nature, and all these various modes of expressing that conviction not only improve the intelligibility of his studies but help to reveal the fundamental nature of his philosophy and to peel back the layers of reticence that often mask the nature of Gilbert himself.

The Society of a Geologist

During his years with the Powell Survey, Gilbert sculptured his scientific thought, both as a method and as a set of working concepts, into a shape he permanently retained. During this same period, he similarly refined his social habits and personality. In a peculiar fashion, these various dimensions to his life came to display certain affinities.

In 1875, shortly after he returned from a field season on the High Plateaus, Fannie gave birth to a daughter, whom they named Betsy Bent Gilbert. About the same time, his brother-in-law, Archibald Marvine, became seriously ill. Despite constant care (Gilbert spent many nights as a nurse), Marvine died in March 1876. The shock was double: Gilbert lost not only a close friend and relative but a brilliant scientific colleague whose opinions on western geology were at least as stimulating as those he received from Powell. Indeed, both he and Powell wrote obituary notices. When a son was born to Karl and Fannie on December 6 of that year, he was named Archibald Marvine Gilbert.

The Gilberts lived in Le Droit Park, a suburb frequented by survey members. GK's range of social functions practically mirrored his scientific associations. The most frequented and lively of his numerous scientific clubs, for many of which he served in administrative roles, were the Philosophical Society and, after 1879, the Cosmos Club. He valued these associations quite highly—except for card games with Powell, Henshaw, and the like, they provided most of his entertainment away from home. He rarely attended theaters and eventually retired from them so much that he would buy tickets for Henshaw to escort Fannie to a show while he stayed home. His social life was circumscribed in revealing ways. If he did attend a play, he avoided sentimental melodramas, for he could break down into weeping. The same could happen at his evening readings. At this point, "unable to go on," he would "temporarily relinquish the book to another," Henshaw recalled. It was as though he simply let Powell act out any public or emotional displays he himself might have wished for.[51]

Most of Gilbert's evenings, as had been the case in his youth, revolved around the home. He read a great deal; and frequently, when he invited guests over for an evening, he read aloud from favorite books—a perennial selection was Clarence King's *Mountaineering in the Sierra Nevada*. For the most part, his scale of living was modest, and he commonly helped with domestic chores, including mending his own clothes. In particular, he kept the family financial accounts conscientiously. In one respect he had to: his salary was not great, and often it was simply deferred. In 1878 he entered into his diary: "Powell still owes unpaid salary, $217.50." A short time later, he added: "Major Powell requests me to wait until May 1st before receiving more of my pay. I have now received all of my salary for 1877 except $5.00 and none for 1878." In another sense, however, the accounts he kept on his personal affairs were not much different from those he kept on nature. In both cases, he was never satisfied until they squared. Nature, like certain geologists, was expected to balance its books.[52]

Those who knew him well spoke often about this "philosophy of self-control," which they greatly admired. "As deeply as certain times of unhappiness were impressed upon Gilbert's self," recalled William Morris Davis,

> frankly as they were spoken of to a very few, they were never made known to the greater number of his associates and they must now lie buried with him; buried all the deeper because

his courageous philosophy of life led him to live joyously. He kept his griefs and disappointments to himself and radiated only good cheer upon his comrades.[53]

In probably the best description by a contemporary, W. C. Mendenhall put it this way:

> In sheer balanced mental power, Gilbert was probably unsurpassed by any geologist of his time. Fundamental among the qualities of his mind were self-knowledge and self-control. These qualities he possessed in a degree equalled by few. That mind which he knew and controlled so well was a quiet, efficient, powerful instrument, which functioned perfectly. Thus he was the very antithesis of the brilliant, temperamental, erratic genius. He recognized both his powers and his limitations, and did not undertake that which he was not equipped to do.[54]

There is something awesome about Gilbert's self-containment, something gyroscopic. Undoubtedly his preference for expressing strong emotions only in the context of small groups of close friends simply reflected the circumstances of the Nutshell and his tendency to re-create it whenever possible. It was an inclination aggravated by the death of Bessie at age seven, a trauma he never recovered from. Yet throughout all this it is as though his personality were taking on the same order he saw in nature, or perhaps his perception of nature was a projection of that personality. No matter which, for either way the shape of his life came more and more to embody the same rhythms and equilibrium he read across the face of nature.

Gilbert's temperament differed from that of his contemporaries in much the same way his scientific theories did. His story about an incident that occurred on the Wheeler Survey in the vicinity of Bryce Canyon illustrates this point. "I rode along a high plateau in southern Utah," he reminisced. "My companions were Hoxie, a young Army officer; Weiss, a veteran topographer, who mapped our route as we went; and Kipp, an assistant whose primary duty was to carry a barometer." Suddenly, the ground fell away before them, and they stood on a cliff "spellbound" by the sweeping vision. Kipp broke the silence by exclaiming: "Well, we're nicely caught!"—with the result that, as GK put it, their "tense emotion found first expression in a laugh."[55]

When John Wesley Powell wrote about that country, he portrayed it as a scene of heroic splendor and almost melodramatic natural majesty. When Clarence King triumphantly planted himself atop Mount Lassen, he surrendered to the expansiveness of the view. "What," he exclaimed, "would Ruskin say if he could only see this!" When Clarence Dutton stood on the brink of the High Plateaus, he rhapsodically claimed that the geologist became a poet, as he peered toward an eastern horizon of sublime desolation. But for Gilbert every theory of nature had a counterforce—whether as an intellectual assumption, a natural process, or a Kipp—which worked to contain his enthusiasm, to balance his perception, and to discipline his understanding into a frame of equilibrium. Like that natural world he presented, Gilbert never worked to excess.

The *Henry Mountains* made its author an international figure in the earth sciences. His 1875 report for Wheeler had established his credentials as a scientist and an explorer; the *Henry Mountains* established him as a theoretician. And there was still more to come from his years with the Major. Some scrupulous measurements on the fluctuations of the Great Salt Lake and an analysis of it in terms of meteorological physics marked another milestone on the long road to *Lake Bonneville*. At the same time, Gilbert labored conscientiously as an editor. In 1878 Henry Newton, a colleague from the Ohio Survey, tragically died during a round of fieldwork for the federal government at the Black Hills of South Dakota. All he left were some notes and a rough, penciled manuscript, which Gilbert edited for publication. In addition, he virtually wrote one important chapter himself from Newton's cards of raw data. It bore Newton's title, "Structure and Age of the Black Hills," but it offered an analysis developed from work on the Colorado Plateau. The mountains were the relic of a dome or arch; its erosion followed according to a mixture of Gilbertian laws and Powellian classifications. As Emmanuel de Margerie wrote Gilbert years later from Paris, "most of what may present any general interest in Newton and Jenny's description of the Black Hills of South Dakota ought to bear your signature."[56]

It was appropriate that Gilbert should conclude his work under Powell with a study of structure and drainage and a project that put him into a position of editor rather than investigator. Though they used radically different language, the question of antecedent drainage was a favorite theme of both men during this period. On the horizon, however, were a position with the U.S. Geological Survey and *Lake Bonneville*. The sequence and associations are suggestive. During these years, Gilbert's creativity had represented a point of acceleration and concentration; relative to what followed, it resembled the motion of a river through a narrow gorge before debouching listlessly into the delta of a great lake.

4. A Great Engine of Research

*His first work under the present Survey, the Lake Bonneville monograph,
still retains its place as the premier among the more than 1,300 scientific
and technical volumes of the United States Geological Survey.*
　　　　　　　　　　　　　　　　　　　　　　　—George Otis Smith

I have the honor to request, for Mr. Russell's use, a revolving chair . . .
　　　　　　　　　　　　　　　　　　　　　　—Grove Karl Gilbert

The Division of the Great Basin

Its first director, Clarence King, fielded the U.S. Geological Survey with the same vigor that, at age twenty-five, he had applied to the Exploration of the 40th Parallel. Graduate of Yale's Sheffield Scientific School, celebrated mountaineer and author, explorer, geologist, and gifted raconteur, King seemed an ideal candidate for the directorship. The exceptional quality of his 40th parallel reports had made good his claim "to give to this work a finish which will place it on an equal footing with the best European publications, and those few which have redeemed the wavering reputation of our American investigators." Through them, as William Goetzmann has remarked, he "incorporated the West into the realm of academic science."[1]

King proposed to extend the same standards to the national survey—as a scientific institution, it would be placed on a par with the new scientific and technological schools mushrooming across the country on the German university model. He hired some of the seasoned professionals he had known on the 40th parallel survey and retained most of the chief geologists who were working on the separate federal surveys; in fact, as in Gilbert's case, they persisted in their old projects.

In many ways the old survey patterns endured, but they lacked the competition which had marred them for a decade. King carved the West up into divisions, each with a chief geologist and his staff. In effect, the national organization sponsored four small surveys with the Washington office acting as a disbursing agent and a congressional liaison. While King seemed only to project his past field experience into the novel circumstances he found in 1879, he had plenty of new ideas too: he wanted to rationalize and upgrade the whole enterprise. This meant advanced scientific research and its almost inevitable political companion, resource conservation. In contrast to every other "intelligent nation," King wrote, "today no one knows, with the slightest approach to accuracy, the status of the mineral industry, either technically, as regards the progress and development making in methods, or statistically, as regards the sources, amounts, and valuations of the various productions." "The epoch of the pioneer is practically passed," he warned; the real problem of the West was "industrial," to bring the population into "equilibrium with local resources." To this end, he concluded that "the intention of Congress was to begin a rigid scientific classification of lands of the national domain" and "to produce a series of land maps" which could serve Congress and the populace as a basis for rational decisions.[2]

But what really fascinated King was mountains. He had made his reputation, both public and scientific, by climbing, celebrating, and explaining them. The romantic and the scientist in him knotted together in his dramatic account of a first ascent of a High Sierra peak. Ringing his hammer on the topmost rock, before surrendering to the sublimity of its panorama, he proclaimed it Mount Tyndall, after the British physicist. Such exploits, moreover, combined with a prodigious social charm: King moved in clubs as brilliantly as he did along the Sierra crest line. In 1871, he scrambled to the summit of the highest peak in the range and the tallest in the continental United States. The man who scaled Mount Whitney and who subsequently assumed the directorship of the USGS appeared to such contemporaries as Henry Adams to occupy the summit of a civilization.[3]

Naturally the problems in geology that most fascinated King were those connected with mountains—orogeny and ore formation. While on the summit he might quote Ruskin, in the mine shaft he turned to physics and chemistry. Metallurgy, especially, was a source of technical information. As a consequence, he established a chemical laboratory at Denver and another—many of whose instruments he financed out of his own pocket—in New Haven. He hired

Carl Barus, a recent Ph.D. from Würzburg, and William Hallock, from Columbia, to staff the latter and to conduct experiments on the geophysical properties of rocks. In his traverse of the Great Basin with the 40th parallel survey, King had worked out a sequence of wet and dry epochs from the chemistry of sediments deposited in the Pleistocene Lake Lahontan in Nevada. He wanted to address some of the larger questions of geology through the same methods.

From his new platform, King campaigned for mining reform. He was appalled by the waste, ignorance, and speculation rampant in American mining. Considering mining rather than agriculture as the essential component of western progress, he sought to organize the business more efficiently—through scientific research and statistical data. At any rate, his emphasis on economic geology won the applause of Congress, particularly the western contingent.

The first stage in the program, in addition to the laboratories established at Denver and New Haven, was to send competent geologists to the mining regions and to compile statistical inventories of mineral resources. King dispatched his old hands, Samuel Emmons and James Hague, for this assignment and added Raphael Pumpelly and George Becker, a brilliant Heidelberg Ph.D. and graduate of the Royal Academy of Mines.

The members of Powell's survey were left to continue their old projects. Dutton completed his manuscript on the High Plateaus, then toured the Grand Canyon in the company of William Holmes in preparation for the *Tertiary History*. Gilbert, meanwhile, was placed in charge of the Division of the Great Basin. He had no office, an uncertain budget (temporarily funded by $2,000 from King's personal accounts), a staff only on paper. But he had something more important—a magisterial topic that both he and the director were eager to investigate: the relic terrane of an enormous Pleistocene lake which he had named Bonneville in 1872.

The broad history of Lake Bonneville was known. It had been publicized by scores of trappers, beginning with Jim Bridger, and by army explorers, on the example of John Frémont, who supplied the name Great Basin for its broader hydrological unit. W. R. Blake, J. D. Whitney, and F. V. Hayden had all commented on the region. Between what King had deduced from mineralogical considerations of Lake Lahontan and what Gilbert had reckoned from structural and geomorphological evidence gathered during the Wheeler Survey, the story of Lake Bonneville was that of a basin which had periodically flooded and drained, eventually leaving only a vestigial relic, the Great Salt Lake. Sediments, alternately of lakes and of alluvial fans, told the general story, while the shorelines etched into the Wasatch related the most recent episode. In 1879 only the broadest themes to the history were known, founded on the physical evidence of scattered bars, deltas, and terraces and on the mineralogical evidence of intercalated layers of evaporites and clays. GK directed his research toward specific questions: the processes of bar formation, the location of an outlet, the measuring of bars, once level but now warped, a quantitative study of changes in water level, and so on.

Gilbert had suggested in an annual report for Wheeler that the oscillations in the flooding of the lake correlated with the oscillations of glaciers, that "the Bonneville epoch was synchronous with the glacial epoch." Both represented "secular cycles of climate." This, in turn, had clear political and economic meaning. "The history of Bonneville," he wrote, "is therefore the history of the ancient climate of Utah, and is thereby closely linked to the material interests of the Territory." Agricultural settlements had already matured along the ancient shorelines, and the "problem of secular change" in climate was "of such vital importance to the agriculture of an arid domain" that, he pointed out, "the public domain presents no more important problem to the survey." The investigation of Lake Bonneville involved theoretical work into meteorology which translated into geosocial meaning. That appealed to Gilbert.[4]

Yet he discovered through his preliminary fieldwork that "the magnitude of the subject had been underrated." For several years he continued to underrate it. Nevertheless, in 1879, the fieldwork moved briskly. The major themes were so fixed in his mind that, as with the Henry Mountain research, he developed nothing novel "but merely added details of configuration and illustrations of old ideas."[5]

With the coming of fog and snow, it became obvious that Gilbert could not complete his examination in a single season, even by prolonging it into December. A few days after Christmas, he wrote King in disappointment: "The subject grows—so that I have now nearly as much work laid out as when I came to Utah." No matter what his desires were, the weather stalled any further work. "The days are short and storms encroach seriously on one's time. The grass is dead and the forage question hampers all movements." Besides, as he wrote earlier, "winter work is poor in quality. Exposures are often hidden by snow. Numb fingers make poor maps and sketches. A biting wind distracts the mind more than the broiling sun. Finally there are three localities I think it important to see before going to press—which are now buried under snow and likely to remain so till spring." He returned to Washington to confer with King and to await an early spring thaw.[6]

In 1880, with a larger force, Gilbert formed two parties—one for mapping under Gilbert Thompson, another, which he joined, for geology. The second group included I. C. Russell as a young assistant. Supplies were shipped by rail to strategic rendezvous, while each party was outfitted with "a four-mule wagon, three pack mules, and a quota of saddle mules, and upon the wagons were loaded the paraphernalia of camp life." Fifteen men in all, the parties moved south to Utah Lake, then southwest to Sevier Lake and a complex system of mountains and shorelines which, though united by large channels, separated the southern from the main body of Lake Bonneville. Meanwhile GK dispatched Russell to reexamine the outlet discovered at Red Rock Pass in the north. The climate made work difficult. Springs were few, water was precious, and frequently the parties had to supply themselves with "water from tanks maintained by the railroad." In the late fall, snow halted operations, much to Gilbert's disappointment. By November 13, the field season had folded.[7]

Gilbert had to establish a regional office in Salt Lake City, too, during these first years. There was considerable confusion over procedures, proper vouchers, official titles. At one point, in mild exasperation, he wrote King: "Permit me to call attention to the fact that, while I have by invoice and purchase acquired considerable office furniture, I have as yet no official authority to hire an office." He settled the problem soon enough by "hiring an entire building and with it a yard and a barn." Requests for vouchers, rubber bands, and revolving chairs soon followed. Calling attention to the fact that there were a variety of brands on the government mules, he respectfully suggested "the propriety of the adoption by the director of a uniform pattern." By pooling their private collections of scientific books, the geologists of the Great Basin Division overcame the deficits of a "frontier town." Gilbert implemented a thorough system of cataloging "by means of cards."[8]

Somehow he sustained his scientific work. Field notes were elaborated; gauges for measuring the fluctuation of the Great Salt Lake were installed; experiments on the sedimentation of evaporites were begun; and he completed his careful essay on hypsometry, a clear jewel of logic and conciseness. Meanwhile he farmed out much of his text on Bonneville—sending his field drawings to William Holmes for engraving, publishing a preliminary essay, and confidently predicting only another year to finish the full monograph. As he entered the 1881 field season, however, he regarded his study of Bonneville mainly as "a first chapter of the study of the group of continental basins known as the Great Basin." He projected half a dozen investigations into such themes as sedimentation processes, shore processes, and crustal deformation, each with its own memoir.[9]

He never realized that research program, for a few months earlier Clarence King had resigned as director. King had been the guiding genius of Bonneville—in a sense, its collaborator. His resignation left Gilbert without a further field season in which to conclude his study before entering an administrative position as an adjutant to King's successor, Major Powell.

King's meteoric career now became a painfully visible falling star. His interest in mining became less theoretical and more practical; he gambled on mining speculations in Mexico and lost. Except for a short article in 1893, his scientific contributions ended, and financial necessity forced him into a career as a mining engineer. Torn between a sense of scientific duty and a romantic primitivism, a schism he was unable to reconcile, he entered into a clandestine marriage with a black woman, suffered a nervous breakdown after the Panic of 1893, and finally endured a pathetic death from tuberculosis in 1901. It was an experience strangely predicted by his relationship to his beloved Sierras, the segregation of those forces he united atop Mount Tyndall. The man who stood on Mount Whitney in 1871, having scaled the summit of the American natural and social landscape, learned a year later that he had accidentally climbed the wrong peak. Instead of Mount Whitney, "the best and brightest man of his generation"—as John Hay described him—had climbed a nearby peak by mistake and left a memorial there. That misplaced summit symbolizes King's tragic career.

The Revolving Chair: Chief Geologist of the U.S. Geological Survey

When King left the Survey, it was a going concern—ambitious, growing in talent and scope, its first major crop of monographs ready for harvest. Powell stepped in, as he had with Wheeler, in time to reap the reward. Cautiously at first, then swiftly, the Survey altered its direction and organization under his touch. According to Powell, with King there had been too many Germans, too many Ph.D.'s, and research, despite a commendable program of theoretical inquiries, had been too closely allied with industrial capitalism.

Powell reversed those trends. The U.S. Geological Survey under the King and Powell administrations would differ exactly as the Exploration of the 40th Parallel had from the Geological and Geographical Survey of the Rocky Mountain Region. Just as King had extended the studies and promoted the personnel he had known from the 40th parallel, so Powell continued his own fascination with natural history while he elevated his former associates, among them Gilbert. Instead of an industrial science, however, Powell promoted an agrarian one. Instead of subsurface geology, he emphasized a geology pertinent to topography. In place of geophysics and geochemistry as instruments to unravel earth structure, he reinstated paleontology in an effort to decipher earth history. The Survey became a major center for paleontological research, with thick volumes of plates and descriptive text clattering off the government presses. The study of ancient fish, extinct ferns, and primitive peoples preoccupied the director and served by analogy to underwrite problems in the human settlement of the West. All of them belonged to the general topic of natural history and could be summarized on maps as natural and cultural topography. It is no accident that Powell's intellectual alter ego, Lester Frank Ward, directed the Survey's Division of Paleontology.

In short, where King emphasized the conservation of minerals, organized by statistics, Powell campaigned for reform in land and water resources, as expressed in a topographic map. Their political reforms thus coincided with their theoretical interests in geology. King was especially intrigued by mountains, Powell by rivers. Powell was interested less in the geophysics of Lake Bonneville than in its geopolitics; he no more desired a technical, industrial science than he did an industrial social order. Where King looked to the German university and research institute for his model, Powell preferred the native land-grant colleges and homegrown, often self-educated talent. His was a Populist science.[10]

Such a science relied on paleontology as physics relied on mathematics. Under Powell's administration paleontology flourished, at times scandalously so. Even before the consolidation of the western surveys, rivals like O. C. Marsh and E. D. Cope began hijacking specimens and plundering the public domain for its fossils, resembling nothing so much as brawling robber barons thieving the public lands of fossil fuels. The spectacle was tolerable only because, in Dutton's words, "the geologist seeks for facts in order to learn geological history and causation—in short evolution." The bones were the necessary facts.[11]

As well as redefining its mission, Powell restructured the Survey administratively. Basically he centralized it. Scrapping the regional offices in their existing forms as miniatures of the Washington office, he redesigned the subjects according to geologic topics rather than regions. To pursue certain topics, the Survey might install subordinate regional offices. The reason for the change was simple: the senior geologists were being lost to administrative chores. In many cases the new strategy worked; in Gilbert's case, however, it removed him from sustained research for a dozen years.

Powell had directed the Bureau of Ethnology since 1879; he held the post concurrently with the directorship of the Geological Survey. Compounding these tasks with his heady expansion programs for the Survey and with his increasing agitation for land reform drained his time. While he outlined strategies, he needed someone to arrange his administrative fossils. His loyal friend got the call. So in 1881, while still directing the lame-duck Division of the Great Basin from Washington, Gilbert dispatched I. C. Russell to Lake Lahontan in Nevada to produce a companion study to the much delayed *Lake Bonneville*. His own time was spent in revising his reports and "in various duties connected with the general work of the office." That phrase was to be repeated with saddening regularity for the next decade.[12]

This is not to say that GK's labors in the national office were trivial. In terms of the administration of the Survey, they were not. What was lamentable about his tour of duty was that it deprived him of prolonged research opportunities; it turned his famous equanimity of judgment from adjudicating geological problems into administrating geologists. But, as Powell's closest adviser, he contributed immeasurably to the character and practical machinery of the new survey as it became, in his own words, a scientific trust. Even to Harvard observers it was obvious that "an important share of his thought was represented in the plans and assignments announced in administrative reports over the director's name instead of his own." Henry Henshaw noted that even in decisions involving the Bureau of Ethnology Powell habitually relied on Gilbert's sound judgment.[13]

Until 1883, Gilbert continued to supervise the Great Basin studies with a staff including W J McGee and Willard Johnson, in addition to Russell. Except for a reconnaissance trip to Bonneville and the eastern front of the Sierras, where he rendezvoused with Russell for a few weeks and inspected the fault line of the 1872 Lone Pine earthquake, he managed the division from Washington. His Bonneville monograph meanwhile languished—eventually, before its 1890 publication, most of it was brought out piecemeal in the annual reports like a scientific serial. But, if the monograph on the lake stagnated, the investigation of the Great Basin evaporated. Despite his loyalty to Powell, GK withdrew from this promising field with remorse. In his annual report for 1883–84, he wrote the director:

> While I recognize fully the considerations which led to the closing of this investigation of the Great Basin, and while the wisdom of your decision is unquestioned, I yet find myself unable to lay the work aside without the tribute of regret and expression of a hope that it may some day be resumed by another if not by myself.[14]

He turned his notes over to his assistants, chiefly Russell. That, too, was an omen of the years to come.

In 1884 Powell transferred Gilbert to the Appalachian Division, again as geologist-in-charge. The field was new to him, the existing literature large. Gilbert scanned the unfamiliar terrain for three weeks in 1884, touring the mountains of North Carolina, Tennessee, and Georgia. He organized the geologic work of the division as sensibly as he could. To supplement the field research, he initiated an extensive bibliography, which swelled to eleven thousand cards in a year. He reserved for himself, however, a study of the relic stream terraces in the mountains. Later in the summer, he twice visited New England and New York—again to examine terraces and "evidences of post-glacial movements of the earth's crust." "I will personally examine the terrace system of the Atlantic drainage," he earlier wrote Powell, "with the intention of ultimately correlating the terraces with the levelling deposits of the coast."[15] This bootlegged fieldwork represented an attempt to transfer the lessons and topics of Lake Bonneville to its eastern equivalents—the Great Lakes and the Atlantic seaboard. This fractured inquiry, continued for several years, focused especially on Niagara Falls, which took Gilbert pretty nearly outside the Appalachian Division but close to Rochester.

Unfortunately, he lacked the opportunity for much more. He never acquired a large party of assistants, such as King had provided in 1880. He did most of the investigation alone—occasionally he rendezvoused with friends for brief interludes, but the bulk of his research was solo. He relied on engineering surveys of Niagara and the Great Lakes whenever possible, as well as previous geological surveys—a luxury unknown in the Bonneville salt flats or the House Range. The thick veneer of glacial till, the heavy vegetation, and the alteration of the landscape by human labor—all complicated the study.

In the end, as happened with Russell and Lake Lahontan, Gilbert simply handed his notes and observations to others, such as Frank Bursley Taylor, or urged others to proceed with the research he lacked the opportunity to pursue, as happened with the experimental analysis of Appalachian structure conducted by Bailey Willis. He accepted the outcomes philosophically: "As I do not believe in the establishment of scientific preserves I have no complaints to make and only a shade of regret that I am not in it; otherwise, I am proud of the way the work is being done." Taylor and Willis were equally generous. Taylor named an arm of the Pleistocene Great Lakes that had lapped over Rochester the Gilbert Gulf to honor the aid he had received, and Willis memorialized Gilbert in the preface to his *Geologic Structures*: "To G. K. Gilbert, who thirty-eight years ago suggested the study of the mechanical principles underlying structural geology, I owe my introduction to the subject. 'The Mechanics of Appalachian Structure' was inspired by him."[16]

In 1884, Gilbert and Dutton helped Powell escort a group of scientific dignitaries on a tour of the Far Southwest. The Zuni pueblos were one item on their agenda, and not surprisingly, considering its role in their careers, the Grand Canyon was another. After 1884, as he progressively served as the official confidant of Powell, as his equable judgment was sought more and more on problems involving maps, principles of correlation, and administrative trivia, GK grudgingly surrendered even more of his field research. His summer excursions now resembled working vacations more than the field trips he had known in Utah; fieldwork had to be left to his assistants. If Wheeler had bound him to military parties dedicated to topographic mapping, tantalizingly dangling whole mountains before him, the Survey under Powell's leadership disheartened him no less by its bureaucratic regimen and its own monomania for topographic maps, which compelled him to look at mountains he had little interest in. Each month his dolorous report announced that he still worked to prepare his Bonneville monograph for the press, and each month he lamented that office duties preoccupied him.

By 1885 the tenor of office work had Dutton, though a little precipitously, forecasting the Survey's ossification:

Our Survey is now at its zenith and I prophesy its decline. The "organization" is rapidly "perfecting," i.e., more clerks, more rules, more red tape, less freedom of movement, less discretion on the part of the geologists and less out-turn of scientific

products. This is inevitable. It is the law of nature and can no
more be stopped than the growth and decadence of the human
body.[17]

There were a few breaks in Gilbert's routine, all of them working vacations. His glacial studies in New England and New York, as
well as his 1884 tour of the Far Southwest, belong in this category.
In 1888, having brought the insights of the West to bear on eastern
geology, he carried them across the Atlantic as he journeyed to
London to attend the Fourth International Geological Congress. Although he represented the United States and performed some light-
hearted espionage for the Geological Survey, he financed the trip entirely out of his own pocket; it was really an institutional version of
his fieldwork. It is interesting to observe that once again, as with his
social clubs, he required a formal, scientific context before he participated. Not that he behaved stiffly—he wrote that it was a jolly
trip. He met a concourse of British luminaries and nobles but, like a
true innocent abroad, he never lost his Yankee baggage, finding in
Europe's splendors occasions for awe and scorn equally. The *Henry
Mountains* was his calling card; he needed no further introduction.
He attended the British Association for the Advancement of Science
meeting as well as the international congress, meeting Archibald
Geikie, Edward Burnett Tylor, and Baron von Richthofen, among
others. The excursions sponsored by the British association impressed him with their planning and conduct; in comparison, American versions were "chaotic affairs, without adequate provision and
without guidance." Thus, when he had the opportunity in 1891 to
organize a cross-country American tour by the international congress, GK relied on the lessons he had learned in Britain. In determining the sleeping arrangements on the train, for example, he decided: "We shall herd the Germans at one end and the French at the
other and interpose a dining car in the middle of the train to pull
them still further apart."[18]
    But the British affairs could equally bog down in a tar pit of con-
vention. "The giving of thanks is a flagrant example," he wrote a
friend:

The chairman makes a speech and introduces the lecturer; the
lecturer gives his lecture; the chairman makes a little speech
and introduces Mr. A.; Mr. A. makes a speech and moves a
vote of thanks; the chairman introduces Mr. B.; Mr. B makes a
speech and seconds the motion; the chairman puts the motion; the lecturer makes a speech to thank the audience for

thanking him and Mr. A. and Mr. B. for moving the vote; the chairman dismisses the audience with another speech if he can think of anything to say.[19]

He was wined and dined nobly. For one who was used to the tenor of the Nutshell or the companionship of a mule on the Bonneville salt flats, it was all "quite overpowering . . . No American city ever boarded, lodged, wined and carriaged me for a week—or if it did I am not going to tell of it." "Am I not a fortunate Yankee," he asked, "to have this glimpse of high life dropped onto my path?" But, as if he wasn't quite certain, he added: "I am sure that I am enriched by the experience." In the end he couldn't refrain from chiding British formality, however, in either its social or its architectural expressions. "A town street," he observed, "is a box-canyon with occasional doorways through the walls . . . the habit of barring out the wayfarer is a mere survival of lawless times many centuries ago, and our conservative cousins are simply slow to discover that they have become civilized." The choice between the baronial lodgings at the estate of General Pitt-Rivers and a buckboard of straw at Toquerville, Utah, ended in a draw.[20]

During the London meetings, Gilbert had access to the Athenaeum Club, which he found about as hospitable as a glacial crevasse. After the meetings adjourned, he toured the British Isles, pausing at sites of geological notoriety like the Parallel Roads. Then he crossed the channel to France. He stayed as the guest of Emmanuel de Margerie, and this visit consolidated a lifelong friendship between them. Through de Margerie he met Colonel de la Noé—to everyone's delight, for Gilbert's essay on landform processes in the *Henry Mountains* had inspired much in de la Noé and de Margerie's own volume on landforms. (Incidentally, he picked up some valuable tips on mapmaking from the French army engineers.) But the Paris scene prodded him into an outburst that may stand as an epitaph to his sole venture to Europe. When his hosts proudly pointed out the construction of the Arc de Triomphe and the Cathédrale de la Sacré-Coeur, the Yankee, classicist, and engineer in him rebelled. He could only exclaim: "What a worthless use of money!"[21]

When he returned to America, Gilbert found his administrative burdens increased. For Powell, as for King, other interests were supplanting geology. By 1888 Powell had effectively transformed his concern with geology into a program of land reform and agrarian conservation, which took shape as the Irrigation Survey within the U.S. Geological Survey, and he offered Gilbert the post of director of the new division. It was an attractive invitation. His "western fever," as Gilbert called it, still burned strong at age forty-five, and the practical consequences of the program as well as its emphasis on engineering were powerful inducements. It could amount to a theoretical and quantitative continuation of his fluvial studies at the Henry Mountains. But *Lake Bonneville* had yet not found its way to the printer, and Gilbert still hoped to expand his Great Lakes studies into an eastern version of his Bonneville studies. There were also personal reasons: Fannie had become an invalid, and his two adolescent sons needed more attention. He declined the job, and it went to Dutton. Instead, Powell promoted Gilbert to the post of chief geologist. That meant that he effectively ran the geologic branch of the Survey, while Dutton handled the irrigation branch, Henry Gannett the topographical, Charles Walcott the paleontological, and Powell the political chores of the Survey and the Bureau of Ethnology.

The Survey opened its throttle and steamed into its headiest period. After a brilliant performance by Powell before the Allison Commission in 1885, it had expanded to the size King had predicted for it, and beyond. It plunged into Powell's pet projects: a topographic map of the entire United States plus agrarian reform through revision in land and water legislation. It seemed that the Survey carved through its opposition like the Green River through the Uintas. But, as far as its chief geologist was concerned, Powell merely added another burden in 1889 when he selected Gilbert to serve as chairman of a committee whose assignment was to standardize map symbols, formations, and correlation procedures. Given GK's skepticism about the value of most geologic chronometers, the choice was ironic, but he shouldered the work demurely. After all, the problem did have significance. The International Geological Congress had designated map and correlation procedures as its primary subjects, the only topics Gilbert addressed that body about, and maps were also the geopolitical focus of Powell's conservation reforms. Thus while Dutton assumed charge of the Irrigation Survey, Gilbert, as chief geologist of the Geological Survey in general, and of the Division of Correlation in particular, undertook the formulation of the scientific foundations of the geological atlases. The arrangement recapitulated the division of assignments which had resulted in the *Arid Lands* report.

Besides, as he wrote Geikie at the time, it offered him an opportunity to check excess enthusiasms. The situation with the U.S. Geological Survey was a microcosm of that existing for geology in general. The American map had its counterpart in the scheme for a geologic map of Europe; the Division of Correlation acted for the first as the international congress did for the other. As Gilbert wrote, there was the danger that both groups would attempt

> to regulate that which should not be regulated by legislation. I had thought very little was likely to be accomplished by it, but I am now convinced that unless the conservatives make themselves heard geology may be saddled with a tyranny of authoritative classification that will seriously hamper its development.[22]

As he had written in an 1887 essay, the problems of taxonomy and correlation merged: "Taxonomy would be conceived by many geologists as an end instead of a means, just as correlation has been conceived, and energy would be wasted in taxonomic refinement and taxonomic controversy." This was a problem of language, and improper language could corrupt facts. Moreover, "there is a tendency of the mind to attach undue weight to classification"; unless such a system was founded on a rational, causal theory of the phenomena in question, the result could be "suicidal."[23]

In particular, Gilbert resisted the effort to universalize the geologic column. Its record, he insisted, was local, not general. This was entirely in keeping with his proposal that the Survey use physical properties to designate a formation, not temporal ones. Where the international congress sought to systematize "the verbal and graphic language of the science," he concurred with it; when it tried to regulate facts, he opposed it. "A classification," he commented,

> if it has any value whatever, is merely a generalized expression of the facts of observation, and is outside the dominion of the voter. If it comprises all the essential facts, its sufficiency will eventually be recognized, whether its authority is individual or collective. If it does not comprise them, it will inevitably be superseded, by whatever authority it may have been instituted. For this reason I am opposed to the classification by the Congress of the sedimentary formations, and likewise to the classification of the volcanic rocks, and I also regard it as ill-advised that the Congress undertook the preparation of a map of Europe, for that—if more than a work of compilation—is a work of classification.

In short, such a body "may regulate the art of the geologist, but it must not attempt to regulate his science."[24] There were only three subjects in geology which required a regulated nomenclature—petrography, paleontology, and stratigraphy. It is interesting that these were precisely the fields Gilbert avoided.

In supervising the publication of correlation essays, GK confronted a similar dilemma. This was an exercise in historical geology. Every stratigraphic system and group were local, and he could not "too strongly and too earnestly insist that a system which is universal is artificial." All this relates, of course, to his conception of geologic history, especially as defined by evolution:

> As in human history there are interrelations and harmonies and a universal progress, but these are perceptible only in the general view; and the student whose preconceptions lead him to exaggerate the harmonies and ignore the discrepancies perverts the meaning of every page.

He rescued himself from the latent evolutionism of this sentence, however, with a peculiar passage which comprises practically his only comment on organic evolution:

> Thus the secular evolution of species combined with the secular and kaleidoscopic revolution of land areas, leads to two antagonistic tendencies, one toward diversity of life of different parts of the globe, the other toward its uniformity. The tendency toward uniformity affords the basis for the correlation of terranes by comparison of fossils; the tendency toward diversity limits the possibilities of correlation.

He might have been describing landforms at the Henry Mountains.[25]

Correlation and taxonomy were important in themselves, but their ultimate value for geologists came with the need to make maps. From the beginning GK had invested considerable time in map questions. When Powell first took over the Survey, Gilbert closely coached him on the policy for a national map; he chaired a committee to review such questions as map colors, nomenclature, and order of publication; and, as editor of the correlation bulletins, he was in an excellent position to continue his influence. He became, in fact, as great a critic of maps as of correlation and taxonomic principles. In Powell's haste, and against Gilbert's firm warnings, many of the early Survey maps were sloppily done; some were merely drawn from earlier hachure maps rather than resurveyed. His criticism had a practical as well as a philosophical dimension: in nearly every one of his field studies under the Survey,

Gilbert had to discard the official map as inadequate or inaccurate and resurvey the terrain himself. In a sense, he was only repeating on a national scale what he had done in his early years on the Powell Survey in Utah. Powell accepted most of Gilbert's proposals for the map and correlation project in principle but continued to support some marginal work in the field. It was on the overly ambitious map program, however, that congressional criticism of the Survey made itself felt. But it was not until the winter of 1902–03 that conceptual, as distinct from practical, difficulties with the map scheme created a need for revisions. When that time came, Gilbert once again served as chairman.

To Gilbert, by the end of the 1880s, it must have seemed as though his career were approaching baselevel. He was a very competent administrator—there is universal consensus on that—but he was a greater researcher. In commending the Geological Survey for its business methods, the Allison Commission should have given much of the credit to Gilbert. Though as an institution the Survey benefited by his promotion, as a science it suffered. As the years passed, the Survey's chief geologist looked more and more like its chief clerk.

At any rate, GK had almost abandoned his own research and shouldered that of others; much of this work was editorial. He treated problems of priority scornfully. To one disgruntled plaintiff, he wrote in prose as brusque as any he ever used:

> In my opinion it makes little difference to the scientific world by whom discoveries are made, and I regard public discussions of questions of authorship and priority as a burden to the literature of science, occupying space and costing energy that could be better devoted. In my own writings I endeavor to give credit to those whose ideas and work I use, but I do not demand that others shall treat my work in the same way, and I do not propose ever to make reclamation of ideas borrowed or observations duplicated by others.[26]

He never did. In fact, his scrupulosity drove him in the other direction. When others pointed out antecedents to his work, he occasionally accused himself of plagiarism in cases that were clearly independent discovery. He corrected his errors publicly on the grounds that it was preferable to amend himself than to suffer someone else to do it for him. As chief geologist, he tried to make this a survey practice.

Correction was Gilbert's assignment as an editor. Normally he

handled the job with tact and sensitivity. In one memorable instance, he returned a manuscript with the genial commentary:
"The manuscript is pervaded by the originality of your amanuensis,
and I fear that our editor, in eliminating that, may fail to attain that
combination of accuracy and grace which would result from your
own careful revision." If there were complaints about his administration, they lay here—in his circumspection about making decisions which involved other people. The impartiality and delib-
eration he brought to geologic problems occasionally aggravated
personnel problems when he tried to transfer the same methods. He
related to geological nature more readily than to human nature: in
his official conduct as in his field research, he worked better alone.
But a chief geologist did not enjoy that option.[27]

He was a superb adjutant, but he made a poor surrogate. In
1889, for example, Powell was unable to attend the annual meeting
of the American Association for the Advancement of Science, but as
retiring president he had prepared an address which he asked Gilbert to deliver for him. The title: "The Evolution of Music from
Dance to Symphony." Both in prose and in speculation it was Powell
at his most effulgent—it is hard to imagine a theme and style more
foreign to Gilbert's own. Though he presented the address deliberately, it must have bordered on parody when he tried to affect a
mood so innately cross-grained. Had he assumed a higher office in
the Survey, something similar might have resulted. To prevent this,
the Survey would have had to change its character radically at a time
when it was in the vanguard of national reform. The regret in
Gilbert's career is not that he failed to achieve the directorship,
though Powell may have offered it to him, but that he made it as far
as chief geologist. Every office like "a Presidency," he wrote, is "one
of the things for which one is twice glad"—once for the honor of its
appointment and once for the relief upon its surrender.[28] Gilbert's
second blessing came in 1892.

Powell's direction of the Survey, especially in matters of western land reform, had earned him enemies, many of them in Congress. When he succeeded in attaching a clause to the Irrigation Act
of 1888 which was interpreted as closing the public domain until
the USGS completed its survey of irrigable lands and reservoir sites,
he gave his critics a common cause. He aggravated the situation further by insisting that the irrigation work necessitated a national topographic map and by bootlegging appropriations to achieve it. He
tried to centralize western settlement as he tried to centralize all
scientific work of the government under a single department. By
1890 he had overreached himself: Congress gutted his map and irri-

gation projects. By 1892, the Survey itself reeled under Draconian retaliations as its budget was cut in half.

Gilbert had remarkably little to do with the political storm that swirled around Powell and the Survey. He left the fighting to the Major. An appearance before congressional committees in 1890, in which he argued that the topographic maps were valuable for engineering and geologic work, was his sole contribution. Instead, he finally nursed *Lake Bonneville* into an almost anticlimactic publication.

Similarly, in 1892, rather than lobbying in Congress, GK made an unsuccessful venture to New England in an effort to secure funds from philanthropists to sponsor a project which became a lifelong dream with him: to bore deeply into the earth's crust in order to make some crucial geophysical determinations. Meanwhile, he revived field research through the only device he could manage—a leave of absence. A year earlier he had begun studying the crater at Coon Butte, Arizona (now Meteor Crater), located east of Flagstaff. When the maelstrom over Survey mapping and conservation practices struck full blast in 1892, Gilbert was elaborating on the meteorlike Coon Butte scenery with evenings at the Naval Observatory telescope in Washington. What time he did not give to the Survey he gave to his personal research. Controversy only marred science; he would no more permit the political currents to halt his observation of the moon than he had let the Colorado River prevent his sighting of Venus from its gorge in 1871. If anything, his philosophical abstraction from the practical affairs of the Survey worked against it. One congressman used the affair as a means of taking the Survey to task: "So useless has the survey become," he huffed, "that one of its most distinguished members has no better way to employ his time than to sit up all night gaping at the moon." Gilbert bore the stigma proudly.[29]

With its appropriations slashed, the Survey painfully restructured itself. Many positions vaporized, including that of chief geologist. The responsibility for notifying terminated employees fell to Gilbert, who did his best to locate positions on state surveys and at universities for those the national survey could no longer support. Powell was badly strained, and his stump of an arm flared painfully; Gilbert consoled him and tried to lessen the administrative burdens. He even foresaw some benefits to the Survey as a scientific institution: "While the Congressional onslaught is disastrous to many individuals, and therefore grieves me greatly, it is not an unmixed blessing for the Survey."[30]

One of those who stood to be blessed was Gilbert. In helping run one great engine of research, the Survey, for over a decade that other engine, Gilbert, had almost stalled as an autonomous and original geologist. Yet the Survey gave as well as got. In a final attempt to rally support, Powell had dispatched him to the High Plains for a study of artesian wells in 1893. This marked the beginning of a return to the West for GK. But administration still had its demands. Though Congress had eliminated the position of chief geologist, Powell asked Gilbert to remain at the post unofficially. He was in that role when Powell resigned the directorship in 1894. In fact, in his faithful labors for his longtime friend, he had come full circle. His final chores found him, in Powell's words, doing "some work on the collections of last summer, especially sorting out, labelling, and transmitting collections pertaining to mineralogy and ethnology."[31]

The Scientist as Aristocrat: George Ferdinand Becker

If Gilbert's relationship with Powell extracted a price from his science, it at least assured him of a scientific job. Though the two might disagree on methodologies and subjects, their friendship lashed their fates together. Others were not always so fortunate. The seasoned corps of theoreticians and engineers King had installed responded to the Powell regime with critical uncertainty and a touch of bitterness. Yet they were as superior as they were different. And no one approximated King's ideal of the academic geologist more than George Ferdinand Becker.

The son of wealthy parents, Becker's education had been exceptional: Harvard, a Ph.D. (summa cum laude) from the University of Heidelberg, graduation from the Royal Academy of Mines in Berlin. Capping it all, he won a certificate for practical work as a "puddler" in the Royal Iron Works. Returning to the United States in 1872, he served two years as a construction engineer at a steel plant in Illinois. In 1874 he accepted an appointment as instructor in mining and metallurgy at the University of California. Here he met King, and in 1879 King offered him a position with the U.S. Geological Survey. Becker not only came but easily convinced King, who shared many of his scientific concerns, to hire physicists and chemists and to open a laboratory. Hence Carl Barus, William Hallock, and W. F. Hildebrand entered federal service—all with Ph.D.'s. Becker projected an ambitious inquiry into geophysics and geochemistry, focusing especially on igneous rock formation and ore genesis. No geologist in the country was better prepared for the task.[32]

When King resigned, he was not—as he had written Becker—a power "behind the throne" who would insure that "your position and interests will be taken care of."[33] Powell substituted his own program for King's, shutting down, for example, the Denver field office with its laboratory. Friction was inevitable, manifesting itself socially as well as intellectually. Powell favored men like Lester Frank Ward and W J McGee, who were largely self-taught naturalists, as was he, and who publicly shared his social-political beliefs. Becker and such men as Samuel Emmons considered this homespun science incompetent. Ward and McGee acquiesced in Powell's belief that geology was basically exploration. But for Becker science came out of academies, not out of boats. Research belonged in formal institutes, not in political reform organizations. Becker's program was no less practical than Powell's maps and irrigation work and no less theoretical than anything the Survey ever published—it was only far more technical. To say that he was simply an economic geologist is no more accurate than to label Willard Gibbs an industrial chemist.

Outwardly relationships in the Survey were cordial, but the schism was never sealed: there were too many cleavage planes. Becker deplored Powell's diversion of the Survey from pure science. He wrote Arnold Hague in 1889: "The Major is head over ears in irrigation problems . . . It looks now as if the Survey would be transformed into a bureau of public works." He and Emmons disliked the Major's mixture of anthropology and geology and the free exchange of staff this involved. Here Gilbert joined them, as Emmons reported to Becker: "Gilbert very particularly suggested that he (Powell) should hire another building and put his ethnologists into it, leaving this for the geologists." This extended to other matters as well: "I asked Gilbert this morning what impression he got that he (Powell) would be likely to do about it. Gilbert said he will think it over for a few days, come to half a dozen minds about it, and then decide suddenly without any reference to what we had said about it." Again, with Emmons as correspondent: "I think that is Powell's word: he always has a word of his own for anything, but as it generally conveys no idea to any head but his own, it is difficult to remember." Becker and his cohorts saw no more point to his programs than to his eccentric vocabulary. They shared his reformist zeal as little as they belonged to the Great Basin Mess, a select group of lunchers in the Survey who clustered around the Powell clique. They were probably appalled at the wager between Powell and McGee as to who had the larger brain, which involved arrangements to have their brains weighed after they died (Powell won).[34]

The pure scientists fought vigorously to keep their old programs, the laboratory, the annual volume of mineral statistics.

There was some success. The Survey even added a superb physicist and mathematician, Robert Simpson Woodward, to the staff in 1884, though he was used predominantly for problems in geodesy and cartography. Most of King's programs were lamed but were not completely destroyed until the debacle in 1892 and the subsequent retrenchment. Becker, Emmons, Barus, Hallock, Hildebrand—all of them went. Woodward had already left in 1890 for an appointment with the Coast and Geodetic Survey. The New Haven laboratory instruments, which had been painstakingly constructed and tested for over ten years, were hauled to the scrap heap by congressional directive; only those King had personally financed remained. However, King and Becker both used Barus' careful measurements to estimate an age for the earth from geophysical theory.[35]

After Powell resigned and appropriations were reinstated in 1894, Becker and Emmons both returned to the Survey under the directorship of Charles Walcott. Becker managed to reinstate his lab in 1900. However, his persistent labors as a geophysicist for the Carnegie Institution of Washington later contributed immeasurably to the foundation of the Carnegie Geophysical Laboratory in 1906, and here he collaborated with van Orstrand in further experimental studies.

Becker was a cosmopolite. He had several times journeyed abroad to inspect foreign mines for comparison with American ones. He had traveled extensively in Germany, Spain, and South Africa. After the cession of the Philippines to the United States in 1898, he joined the official survey of the islands and inadvertently entered several military expeditions, for which he was twice cited for bravery. At home again, his wit and erudition gathered around him a circle reminiscent of that he had known as a child—a circle which had included Oliver Wendell Holmes, Richard Henry Dana, Jeffries Wyman, Louis Agassiz, Benjamin Peirce, and Henry Wadsworth Longfellow.

Although most of Becker's monographs dealt with mining regions, and those mostly in California, the mines existed only to pose an array of theoretical problems. To these he brought an outstanding knowledge of mathematics, physics, chemistry, and geology. Simply as treatises on mining districts, they were rivaled in their time only by Emmons' *Geology and Mining Industry of Leadville, Colorado*. Yet they bristle with dozens of minor technical inquiries, such as the chemistry of kaolinization and the physics of "rock mechanics," that Becker could only touch upon.

"There was no man of his time in the Washington geological world who possessed greater versatility in discussion or such breadth of view," wrote Arthur Day, his memoirist. Indeed, Becker's versatility worked against him: he attacked too many themes at once and pursued few with the rigor he was capable of. Only in a few superb articles—as in his masterpiece, "Finite Homogeneous Strain, Flow, and Rupture of Rocks"—did he apply his full analytical powers. He addressed most of the classic problems in geophysics— such as the age and rigidity of the earth—and he added some others regarding mathematical models of slaty cleavage and the distribution of debris about a volcanic cone. He was the outstanding theoretician of his era, "with the possible exception of Gilbert," Day concluded.[36]

There is curiosity, then, in the relationship of Becker to Gilbert. Socially it did not exist. Becker probably shared with Emmons a certain frustration regarding Gilbert's tenure as chief geologist when he seemed to promote Powell's programs at the expense of economic studies. Some of the prejudice attributed to GK was more imagined than real; most of Becker's and Emmons' scorn was reserved for McGee, and actually Gilbert's research suffered more than theirs. Their social segregation was lamentable, for Gilbert's scientific ambitions intersected at many points with those of Becker. But there was a critical difference in their approach to geology and physics that could no more be reconciled than could the difference between their inherited incomes. That difference had both a social and an intellectual dimension.

Becker was at heart an aristocrat. His scientific idol was Lord Kelvin; he liked the "immortal" Kant (for Gilbert, it was Kelvin the engineer who mattered; for Becker, it was Lord Kelvin, geophysicist). Just as he worked as a "puddler" *after* he completed his doctorate, so he approached geology after a thorough grounding in physics, chemistry, and mathematics. The crucial moment in his early education was the discovery of mathematics. "I was somewhat adrift," he wrote, "when a single lecture on mathematics opened my eyes to the fact that this is a science of great principles and ideas, not a mere jumble of tiresome computations and unrelated Chinese puzzles." He never forgot that lesson.[37]

Consequently, he framed his major theoretical papers in formal, mathematical language and disdained to convert his concepts into simplified, more popular language—precisely the talent Powell had in abundance. Becker was adamant. "The general public of course has no idea what geophysics means," he wrote G. H. Darwin, and he probably placed most of the Geological Survey under the rubric of "general public." In preparing an examination to select an applicant in geology, he wrote Carl Barus: "The examination will be for a 'physical geologist'; but the papers will be purely in physics. This is quite correct because geology is the application of the sciences to the elucidation of the past and present conditions of the earth, and I want a man to apply to that subject matter the sciences of physics."[38]

Gilbert probably agreed but, where Becker was a physicist working in geology, Gilbert was a geologist reaching for physical analogies to untangle his geological riddles. He reached first for geological facts, not for mathematical functions. Like Kelvin, on the other hand, Becker transcribed physics into geology but did not translate physics for geologists. Where Gilbert took special pains to convert formulas and geophysical concepts into a parlance useful to field geologists, Becker defiantly refused to compromise his ambition to set rigorous standards for American geophysicists.

His criteria for those standards identified Becker squarely with the tradition of classical geophysics, a lineage cultivated especially in Britain which included such figures as Kelvin, G. H. Darwin, John Joly, Osmond Fisher, and William Hopkins. But the partnership of classical geology and classical geophysics was, like Becker's relationship to the Survey, an uneasy one. On the one hand, Kelvin's rancorous assault on the time scale of Darwinian evolution challenged the theoretical mechanisms which powered evolutionary geology. On the other, methodological questions and separate techniques created an awkward language barrier between the two camps. In his less conciliatory moods, for example, Becker charged into methodological contests as he had into guerrilla battles in the Philippines. When his theory on slaty cleavage was attacked, he growled: "If I live I'll hammer these geologists till they confess that without a knowledge of mathematical physics they are unfit to theorize."[39] But all the noisy debates between classical geology and classical geophysics tended only to obscure a more fundamental fact: in their philosophical assumptions about the nature of earth history, the two groups were remarkably close. It was the ambition of the best minds of the earth sciences to bring them from proximity into a unified synthesis.

It was appropriate, therefore, that the first textbook in geophysics, Fisher's *Physics of the Earth's Crust*, should be published almost simultaneously with Dutton's *Tertiary History of the Grand Canyon District* and that, paradoxically, the leading American evolutionists—Dana, Dutton, and Le Conte—should be the American authorities Fisher most often referred to. The explanation for the paradox is simple: the Spencerian vision of evolution that underwrites Dutton's history of the Grand Canyon also underpins Fisher's analysis of planetary history. The universal force animating this process was the progressive contraction of the earth. Just as geologists could account for nearly every phenomenon by considering it a fossil and placing it in an evolutionary schema, so geophysicists could relate their subjects to the contraction of the planet. Their chief concerns were thermodynamic—the distribution of energy, as manifested by temperature and pressure, throughout the crust. By assuming a universal compression through time, a host of phenomena could be related mathematically: mountains formed by crustal shortening, the source of volcanic heat, the extrusion of volcanic rocks, gases, and water, the geothermal gradient, and the relative rigidity of the earth. The tendency of the earth to contract and its tendency to evolve were not irreconcilable: the Spencerian formula was one among many that showed how the two histories involved a similar process. In biology this led to a multiplicity of species from a common parent, in geophysics to a variety of earth structures and rocks from an originally undifferentiated mass of earth material. The Reverend Fisher rightly insisted that he had treated his material "in a manner that can hardly offend the most strict disciple of the uniformitarian school."[40]

It was natural that their differences in models and techniques should be magnified when the two groups applied them to the central question of the classical earth sciences—the age of the earth. Though their figures varied, they generally shared two related assumptions: these processes were irreversible and the earth was a closed system. The second was a precondition for the first. Nevertheless, in a striking way the debate between geology and geophysics at the end of the century echoed the debate between theology and geology at its beginning. As Gilbert wryly observed, "Geologists undertook with zeal the revision of their computations, making as earnest an effort for reconciliation as had been made a generation earlier to adjust the elements of Hebrew cosmogony to the facts of geology."[41]

Theology had insisted that the earth was young and that, to fashion it, a series of global interventions of divine origin had been required. Geophysics also concluded that the earth was relatively young and that geologic processes must have evolved at different rates, and probably in different ways, over time. Theology buttressed its arguments with biblical quotations, geophysics with mathematical equations, but the result was identical: an abbreviated age for the earth and, by implication, the improbability of organic evolution by slow, natural processes. In the first debate, geology had agreed with theology that there was historical design evident in the geological record, manifested as a series of progressive stages, but disagreed over the length of time involved and consequently over the mechanisms of causality. In the second debate, geology agreed with geophysics that earth history was continuous and irreversible, whether by evolutionary or by thermodynamic processes, but disagreed again on the length of time being considered and on the methodological techniques involved in calculating it. As Thomas Huxley put it, "Biology takes its time from geology." And geology was determined to make the world safe for evolution. After all, the proposition that outraged Kelvin most about Darwinian evolution was its reliance on chance mutation. Few American geologists would have argued with him on that score. Besides, the numerical differences between the ages arrived at by various geological techniques were not as far from certain geophysical estimates as were other geophysical estimates. Kelvin's figures, however, based on the nebular hypothesis and later work by Hermann von Helmholtz, were as distant from estimates arrived at by other geophysical means, such as tidal retardation, as they were from figures derived from geological assumptions.[42]

Nevertheless, geology and geophysics, as both Gilbert and Becker well knew, were stalemated on a number of intellectual fronts. It had been axiomatic among geologists, for instance, ever since Dana had first broached the topic in 1846, that the continents and ocean basins were permanent features of the crust. Given this skeletal framework, continents had grown at the margins, the oceans had steadily accumulated water, salt, and sediment, and organic evolution had radiated into a great tree of life. To populate these continents, therefore, it was necessary to hypothesize the existence of land bridges to span the rather formidable oceans. The Isthmus of Panama and the Bering Strait were cited as properly uniformitarian examples. Geophysicists, though never a monolithic group, were on the whole skeptical about the supposed land bridges which rose and sank with evolutionary tides, and they were occasionally undecided about the permanence of continents. The continents were distributed too asymmetrically to have been produced by contraction and, except for the soundings brought back by the Challenger Expedition of 1872 to 1876, nothing was known of oceanic geology. After all, the Pacific, as G. H. Darwin suggested, might be only the residual tear left by an escaping moon.

There were still other points of discord. Geologists generally insisted that the earth behaved plastically below its crust; geophysicists, though divided on the question, generally held that the earth was as rigid as steel. To provide a progressive chronology for earth history, geology looked to the remains of organic evolution, denudation rates, or sedimentation rates; geophysics to calculations based on tidal retardation or secular refrigeration or the accumulation of oceanic salts. To furnish the tangential stresses which led to crustal shortening, each invoked a different version of a contracting earth: geology, using the planetesimal hypothesis (after 1905), looked to gravitational collapse; geophysics, using the nebular hypothesis, to thermal cooling. Many geophysicists followed the example of Fisher in accepting contraction while begging the question of its cause. Classical geophysics, in general, allowed only theories which were derivable from the equations of a contracting earth; classical geology, for the most part, accepted only hypotheses inherent in the formulas of progressive evolution.

These were questions that keenly interested the versatile Becker. He fully accepted the geophysical assumptions involved and, given his talent for mathematical physics, he was expertly equipped to comment on such standard topics as the earth's rigidity and age. Not surprisingly, he sided with Kelvin and Darwin. Yet this was not inconsistent with the fact that he wrote a biographical memoir of Dutton. The sibling rivalry between geology and geophysics occurred precisely because they were so related. Gilbert, meanwhile, compromised on the first question and begged the second. He sympathized with physical models and techniques but admitted that, in his "judgment, the weight of the cumulative geologic evidence for the great age of the earth is not counterpoised by the arguments thus far adduced from the physical side of the question."[43] Becker profoundly disagreed. Consequently, while he lived, he gave traditional geophysics a strong theoretical and institutional impetus. When he died, he left a sizable portion of his estate to the Smithsonian to promote that legacy further by funding future geophysical research.

The rift in the Survey ranks reached the proportions it did for reasons that are as much social as intellectual. On both counts this was unfortunate, since it divorced the U.S. Geological Survey from a fertile source of technical expertise, as men like Woodward, Barus, and Emmons drifted into universities, private research institutes, the Coast and Geodetic Survey, or careers as mining engineers. Predictably enough, Becker himself turned away from government science in favor of research institutes, like the Carnegie Geophysical Laboratory, which better approximated German models. Indeed, after both Becker and Woodward had found their way to it, the Carnegie Institution rather than the U.S. Geological Survey became the principal locus of geophysical research. Gilbert acknowledged this fact in 1902 by bringing it his own plans to drill into the earth's crust.

Given that rift, Gilbert occupied uneasy ground. Socially he belonged to Powell's group, clustered about the Great Basin Mess, while intellectually he often came closer to Becker. His education fell intermediately between the two groups—his considerable exploration experience rising upon a college degree in classical subjects. Though he referred occasionally to Becker's works, he turned for mathematical aid to Woodward, who shared his sentiments on the need to package mathematical physics in forms useful to geologists. Consequently, it was Woodward, not Becker, who eulogized Gilbert as "physicist and mathematician."[44]

This schism in the Survey only mirrored the larger quarrel between physics and geology in general, and his uncomfortable position in the Survey in fact only reflected GK's larger dilemma. It brings his peculiar achievement into relief: he was the translator, the liaison between the two camps. What he found objectionable in the physicists' approach was identical to what he protested in the taxonomy of the evolutionists: it attempted to regulate geologic facts. Instead, as he argued by his own example, no topic was so trivial or so refractory that it could not be expressed according to the laws and logic of physics, and no physical law was so inviolate that it could exist meaningfully outside a specific context in the facts of physical geology. What he once wrote about Geikie could be applied equally to himself:

> As in all his writings there is nothing sensational, either in description or in speculation. His inductions are not expanded into brilliant, universal theories, but are modestly advanced with all those limitations which impress themselves on the mind of one who constantly questions nature.[45]

That was the charge. It was the same for the conduct of science as for scientists, but its execution was not equally successful. Gilbert had more success knotting the two disciplines in his geologic monographs than he did their human representatives in the Geological Survey.

Lake Bonneville

It is hard to overestimate *Lake Bonneville*. It climaxed GK's mature science, becoming one of the consummate works of nineteenth-century American geology. Gilbert called it his magnum opus. That methodology which he had pioneered with epigrammatic style in the *Henry Mountains* he amplified and embellished in *Lake Bonneville*. In one quarto volume of four hundred pages, 19,750 square miles of the American continent—much larger than Massachusetts and Connecticut combined—were brought into geological science. The area has never been resurveyed in full.

But the publication of the monograph was also somewhat anticlimactic. Gilbert had first described the lake in 1872 and had furthered his research in 1877, 1879, and 1880; King had described it in 1878. Most western surveys after the Civil War had visited it, as had numerous emissaries of the grand reconnaissance before the war. Under Gilbert's guidance, I. C. Russell had published *Lake Lahontan* in 1885, a study of Bonneville's Pleistocene cousin. At the same time, in fact, memoirs on interior lakes were pouring out of every continent. Besides, Gilbert had already published most of the text in serial fashion over the previous ten years and, with his research program on the Great Lakes gutted by administrative chores, he saw little purpose in recasting his material. In short, the book was stale, a little passé—except for the analysis of postglacial uplift that is its liveliest portion and the topic that most magnetized Gilbert in the future—and publication must have brought more relief than expectation. The principal achievement of *Lake Bonneville* was its comprehensiveness, its codification of lake processes—a function partly that of a textbook. The *Henry Mountains* had been a tour de force; *Lake Bonneville* was a summa.

The theme of *Lake Bonneville* was twofold: "the discovery of the local Pleistocene history and the discovery of the processes by which the changes constituting this history were wrought." The first purpose followed from the second. The monograph consequently opened by describing shoreline processes in what was an axiomatic format. The result of these processes, in various combinations, was a shoreline, and Gilbert took the shorelines of Bonneville for a datum plane or horizon and worked out its Pleistocene history by means of correlations to this horizon. The use of correlations, of course, was standard geologic practice. But Gilbert's correlations did not relate to the stock chronologies of historical geology. There was no stratigraphy, no paleontology (in the usual sense), and no evolutionary summary of the Bonneville region; the baselevel of Powell and the peneplain of Davis were absent. Instead, the correlations were founded on "physical evidence," the most important of which was the record of waves. The effort to correctly correlate the physics of wave processes with the sculpturing of distinctive topographic features is the soul of *Lake Bonneville*. The resulting distinction in the regional landscape between those areas worked over by waves

Bonneville illustrated. Gilbert is shown in the center of the Great Bar at Stockton, Utah, with a plane table, the instrument he used instead of a rock hammer. Drawing by William Holmes. *Reproduced from* Lake Bonneville.

and those unscathed by them furnished the sort of broad, fundamental contrast that Gilbert's imagination loved to seize upon. In his 1875 report for Wheeler, that dichotomy had systematically dissected the differences between the Colorado Plateau and the Basin Range provinces; in *Lake Bonneville*, the differences between a "topography of the lake" and a "topography of the land" were emphasized.[46]

Gilbert had several reasons for insisting on "physical evidence."

For one thing, Bonneville was a relatively recent event. There was no reason to plunge into the mysterious past recorded in fossils because the physical record was still freshly carved into the mountain flanks. For another, the morphological records of early periods, Gilbert insisted, were "to be interpreted only through a knowledge of the processes" that shaped them. "It is through the study of the phenomena of the latest period," he continued, "that the connection between present processes of change and the products of past changes is established."[47] It was to confirm such connections that he had journeyed to the Great Lakes to compare and correlate their shoreline processes to the topographic forms of Bonneville.

These processes, however, were not historical: they functioned not as a dialectic but as the physical laws of mechanics; there was a complete absence of historicity. Gilbert summarized his theory in an 1883 book review:

> Starting from geological agencies as data we may proceed in one direction to the development of geological history, or in another direction to the explanation of terrestrial scenery and topography, and if the development of the earth's history is the peculiar theme of geology, it follows that the explanation of topography, or physiographical geology, is of the nature of an accidental result—a sort of corollary to dynamical geology.[48]

This created a problem of organization—somehow Gilbert had to order geologic events without using narrative history. In *Lake Bonneville* he solved the problem three ways. First, by organizing shoreline processes in the form of a set of axioms, he made it possible to logically, rather than historically, deduce certain topographic forms. Second, by establishing a physical horizon, the shoreline, rather than a fossiliferous or stratigraphic one, he eliminated the need to rely on precisely those tools which led others into evolutionary interpretations of the earth. And, third, he extended the me-·chanical analogy he had designed at the Henry Mountains to explain the Bonneville topography. By equating geological processes with mechanical forces, he was able to organize those processes by the same concept Rankine had used for structures and machines—namely, equilibrium. It is no accident that GK eventually referred to such earth-sculpturing processes as Niagara Falls as a "physiographic engine." To his mind the analogy was very close, and it was entirely applicable to the Bonneville shoreline. Combined with his perception of geologic events as having a plexus of antecedents and consequents and with his belief that geologic processes occurred rhythmically, it insured him against any latent historicism.

The Euclidean format, in short, was essential to Gilbert's understanding of how he should explain the geologic history of Bonneville. Its geologic puzzles could be solved like the problems in Rankine's *Manual of Applied Mechanics*. Geographic terms were established in the form of definitions, processes were arranged as axioms, topographic forms were derived as though theorems to be proven. The proof came by correlating the forms deduced from a rational theory of shoreline processes to the actual features of Lake Bonneville. Moreover, the foundation for such a rational theory rested in fluid mechanics, particularly the study of waves and currents. Gilbert deliberately amalgamated the separate inquiries of engineers, working on harbor modification; geologists, studying coastal deposition and erosion; and physicists, elaborating a theory of waves and hydrodynamics. Each group had its separate literature, and at the time there were few cross references. Consequently, the first quarter of *Lake Bonneville* codified their combined knowledge of shoreline processes in the same systematic, almost Euclidean fashion Gilbert had used in the *Henry Mountains* with regard to stream behavior. Indeed, the two monographs are complementary. In the *Henry Mountains* he had discussed one of the two topographic provinces he found at Bonneville, the topography of the land. In *Lake Bonneville* he would complete the task by describing and contrasting its companion, the topography of the lake. Meanwhile he defended his lengthy compilation by admitting that "the present chapter would have need to demonstrate its raison d'être were it not that the general subject has as yet received no compendious and systematic treatment in the English language."[49]

Thus in one sense Gilbert's contribution was encyclopedic. But in another sense it was creative. "During the progress of the field investigation," he confessed,

> I was unaware of the greater part of the literature mentioned above, having indeed met with but one important paper, that in which Andrews describes the formation of beaches at the head of Lake Michigan, and I was induced by the requirements of my work to develop the philosophy of the subject ab initio. The theories here presented had therefore received approximately their present form and arrangement before they were compared with those of earlier writers. They are thus original without being novel, and their independence gives them confirmatory value so far as they agree with the conclusions of others.[50]

There is no reason to doubt these remarks. Gilbert, who had known E. B. Andrews from the Ohio Survey, quoted his idea that deposition and erosion corresponded to points of deceleration and acceleration in waves and currents. That supplied the fundamental physical analogy which related energy and topographic features for lakes in the same way velocity had related energy and topographic forms for streams. The fields of interest at Bonneville, moreover, were precisely those Gilbert had examined in Ohio. Stylistically and conceptually, the organization of the monograph followed from the *Henry Mountains* as an arrow from a drawn bow. More than merely explicating the laccolith, the *Henry Mountains* had shown Gilbert how to interpret a geologic event generally: it gave him a method. *Lake Bonneville* marked the first extension of that method to a subject GK had been curious about long before he reached Utah.

In the same way Gilbert was able to expand his conception of equilibrium into the new set of processes evident on the shoreline. The resulting profile between lake and shore was identical to that between river and bed, a profile of dynamic equilibrium. Although he did not term it as such, there was clearly a graded beach equivalent to the graded river. Consider this sample of its mechanics from *Lake Bonneville*:

> What may be called a normal profile of the submerged terrace is produced, the parts of which are adjusted to a harmonious interrelation. If some exceptional temporary condition produces abnormal wearing of the outer margin of the terrace, the greater depth of water at that point permits the incoming waves to pass with little impediment and perform their work of erosion upon portions nearer the shore, thus restoring the equilibrium. If exceptional resistance is opposed by the material at the water margin, erosion is there retarded until the submerged terrace has been so reduced as to permit the incoming waves to attack the land with a greater share of unexpended energy. Conversely, if there is a diminution of resistance at the water margin, so as to permit a rapid erosion, the landward recession of that margin causes it to be the less exposed to wave action. Thus the landward wear at the water margin and the downward wear in the several parts of the submerged plateau are adjusted to an interdependent relation.[51]

Gilbert's use of the shoreline as a datum plane was perhaps inevitable. The shoreline furnished the most level feature on the earth's surface, and that appealed to the mathematician in him. In a different sense from that used by Powell, the shoreline was an ideal baselevel. Two shorelines were particularly distinct—an upper one named Bonneville and a much lower one named Provo. While each of the many landforms visible along the old lake could be related to the matrix of processes which had created them, their historical order followed from their position vis-à-vis the Bonneville and Provo shorelines. That is, the forms of shore features could not indicate their place in the evolutionary development of the lake as, say, the fossil bones of *Eohippus* could give stages in the evolution of the horse. Whenever similar landforms existed—as bars, embayments, terraces, deltas, beaches, and cliffs—they indicated a similar matrix of processes but not a similar moment in time or stage of development.

So strict was Gilbert in this regard that he not only dismissed potential fossil equivalents in the landforms of Bonneville but disregarded its actual fossils as well. In his text he benignly recommended an alliance of fossils and physics, though the union was to be founded on common physical evidence at the site. But in a letter to George Becker, who was studying a related problem in Sierra moraines, he wrote:

> The relation of the auriferous gravels to the epoch of glaciation in the Sierra should be made out from the physical relation. I have no faith in the paleontologic determinations from mammals and plants . . . In my judgment there has not been sufficient change in vegetation since the Eocene to give any taxonomic value to a small flora. There is so much variation dependent on horizontal distribution that the variation dependent on vertical distribution is masked.[52]

As though patiently triangulating across the deltas and shore bars, Gilbert carefully elaborated his physical correlations into a physical history of Lake Bonneville. The story involved a closed basin which had been filled with water; the lake basin required processes of both structural geology and climate. Each group of phenomena behaved in a rhythmic way: faulting occurred spasmodically but rhythmically, and climate varied with secular changes that were also rhythmic. The lake resulted when these two sets of rhythms compounded themselves: faulting created closed basins and climate brought greater precipitation.

The basin had at least twice filled and drained. The latest episode had left its record in the shorelines as well as in sediments. Due to considerable oscillations in climate, an irregular staircase of shorelines marked the rise and fall of the old lake. This terraced profile terminated at the Bonneville shoreline, where the lake waters had gradually reached a height sufficient to spill over a mountain pass. The lake's outlet had been an object of intensive search and contention; Gilbert finally identified it with Red Rock Pass, indicating that the lake had drained north into the Snake River. The waters had emptied quickly, since the pass at Red Rock consisted of poorly consolidated sediments. When once the lake crested over it, the rock gave way readily. Accelerating as more and more water poured through, the process continued until the poorly consolidated sediments had perished and left a more resistant bedrock of limestone. This eroded more slowly. In fact, functioning like the spillway to a dam, it worked to maintain a constant water level in the lake— something that would not occur amid the natural rhythms of climate alone. This produced the wide-benched Provo shoreline. Eventually the climate became drier and the remaining water receded to its present diminutive relic, the Great Salt Lake. Both the Bonneville and the Provo shorelines represented the second cycle of the lake; the earlier cycle of lake filling left its record in sediments alone, since its shorelines were reworked and obliterated by the second cycle.

The correlations multiplied. A correspondence was established between the filling and draining of Lake Bonneville and the deposits recorded in the lake beds. Some were indicative of lake sediments (clays and evaporites), others of an arid climate (alluvial fans). The succession of these deposits correlated with climate, which in turn correlated with the shore topography through changes in lake level. Others had suggested such a coincidence, but it was left to Gilbert to elaborate on it. Subsequent correlations depended on this same physical data. An episode of volcanism in the lake (another rhythmic process) was also dated by relation to the shorelines and lake bed deposits.

The most general correlation, however, pertained to climate. Just as Gilbert related sediments and shore forms to the synchronous wave physics of a common lake, so he related glacial activity and the presence of the lake to a common climate. More exactly, he discussed them in relation to physical laws of climate. By an exhaustive analysis, he carefully asserted that a common climate united both phenomena. The series of pluvial lake terraces were analogous to ridges of glacial moraines.

But the drainage and subsequent evaporation of Lake Bonneville's waters did not conclude GK's story. The volume of water in Bonneville had represented a substantial load applied to the earth's crust; its removal, an equal unloading—and the story was incomplete until an equilibrium had been restored. The response of the crust was a question in mechanics of the sort that delighted Gilbert. A geologic literature on this identical topic was emerging in both America and Europe while he proceeded with his own independent investigation. Yet his thoughts have their watershed in the intricacies of the Rochester landscape and the shores of Lake Erie. His favorite baselevel was the geoid—the ideal surface the earth would assume at sea level due to its equilibrium of gravitational forces. In the shoreline strands of Lake Bonneville nature had supplied an analogue. Yet, when measured, the actual shorelines were shown to vary from the horizontal; they were warped, and their warp increased systematically as one moved to the center of the lake. This suggested to Gilbert that the lake floor was warping upward in response to the stress released by the rapid removal of lake waters. He had employed a similar explanation to account for the flexure of ancient shorelines in Ohio.

Others were thinking along similar lines. In particular, Europeans recognized that the rising shorelines of Scandinavia probably resulted from the release of the load imposed by Pleistocene ice sheets. Two years after *Lake Bonneville* was published, DeGreer came to the United States to correlate the postglacial deformation around the Great Lakes—whose reality was by then established by Gilbert—to the Scandinavian phenomenon. Meanwhile, grappling with related problems of load transfers by erosion and deposition, Dutton had coined the term "isostasy" to describe the ideal gravitational equilibrium in the crust—this in a paper delivered to the Philosophical Society in 1889. A year before Dutton's celebrated paper, Gilbert had requested Woodward to solve some questions which involved intricate calculations pertaining to exactly this problem of crustal loading; interestingly enough, Thomas Chamberlin submitted similar requests at the same time. Gilbert was addressing loads induced by water, Chamberlin loads induced by ice. In short, Gilbert's construction was in general again "original without being novel." But, as with his characterization of shore processes, he merited this praise: *Lake Bonneville* was the first major synthesis of the subject.[53]

He proposed three hypotheses to account for the warping and discarded two as quantitatively insufficient. Accepting deformation by loading, he generated a mechanical analogy to the deflection of a beam. In his words,

> For the sake of illustration assume that 129 feet of uplift satisfy the stresses due to 355 feet (129 x 2.75) of the removed water; there remain the stresses due to 645 feet to be satisfied by strains in the crust. Call the basin floor a beam, 120 miles long, supported at the ends, and sustained throughout by flotation so far as its own weight is concerned. Call the modulus of rupture of its material 3,000 pounds to the square inch, and introduce no factor of safety. Consider the beam to be sub-

jected to upward stress by the removal of 645 feet of water from its entire upper surface, and compute by the engineer's formula the depth of beam necessary to stand the strain. It is about 32 miles.[54]

The shorelines posed questions that were both topographical and geophysical, and Gilbert sought to solve them both. The analogy was crude but allowed for the quantitative comparisons he desired. Clearly, the weight of Lake Bonneville was a load sufficient to deform the crust. Yet the Wasatch Mountains adjacent to Bonneville were continuing to grow—that is, the weight of these mountains was not sufficient to depress the earth. This required explanation, and Gilbert multiplied his comparisons. The San Francisco Peaks, for example, represented the relatively sudden deposition of a large volume of volcanics in a manner analogous to the flooding of Bonneville. The upshot was that Gilbert recommended that the crust was neither so plastic as the geologists held nor so rigid as the geophysicists maintained. He suggested that isostatic adjustment was only regional, not local, that rigidity and plasticity were relative terms only. In proving this, he cleverly transformed the surface geology of Lake Bonneville into a tool for exploring the physics of the earth's interior. In that sense, *Lake Bonneville* epitomizes nearly all of his work in geomorphology.

During the long gestation of *Lake Bonneville*, Gilbert submitted a short article to *Nature* magazine which narrated an incident that occurred on a plateau in Utah. "Standing on the verge of a cliff just before sunset," he wrote, "I saw my own shadow and that of the cliff distinctly outlined on the cloud . . . About the head was a bright halo with a diameter several times greater than the head." The spectacle continued "for some hours, the cloud-mass being persistent in position, notwithstanding the fact that its particles had a velocity of twenty-five or thirty miles an hour." He concluded with mock solemnity: "The observation has more than a scientific interest, because, in the popular imagination, the heads of scientific observers are not usually adorned with halos."[55]

Shortly after *Lake Bonneville*'s publication, a reporter from the *New York Tribune* interviewed him and rushed off a lead article about the monograph. It struck GK as "very comic that what I found out years ago should be sent to New York by telegraph instead of mailing the MS. But the reporter never heard of it before, nor the readers of the *Tribune*." They learned soon enough, and, if not to the general public, at least to the fraternity of geologists, *Lake Bon-*

neville entitled Gilbert to stand in that halo he had elsewhere reported.

It is appropriate too that *Lake Bonneville* was published in 1890—the year, as Frederick Jackson Turner pointed out, that the Census Bureau announced that a frontier line of settlement was no longer discernible—for something similar had happened in American geology, in its way a frontier institution. There was a change of phase from exploration to explanation, from discovery to consolidation. *Lake Bonneville* epitomized in one monograph the work of sixty years of exploration in the Great Basin. It was a comprehensive geological interpretation, a codification of processes rather than a catalog of places. It had a touch of the textbook, a faint stammer of pedantry, but it was infused throughout with the excitement and wisdom of science whose conceptual sophistication was still trying to catch up with its accumulated information. "Geology," Gilbert remarked, "is so young a science that a single generation has witnessed a complete revolution." *Lake Bonneville* summarized enough of that revolution that one may grant Gilbert his halo, though one had to take it with the grain of his salt flats.[56]

Lake Bonneville, nonetheless, lacked the literary and conceptual impact of the *Henry Mountains*. Much of its contents had been released prior to its official publication. Like most of its ideas, its prose lacked the crisp, epigrammatic flavor of that in the *Henry Mountains*: it was more massive than incisive. During its long period of gestation, its phrases had been worked and reworked like the pebbles on the Great Bar at Stockton. In some places this gave them a high polish; in others they were eroded almost to minutiae. A little prolix, a little sententious, the passages nevertheless set a rhythm and a style as measured as the foreset beds of a delta. If the phrases were less quotable than those of the *Henry Mountains*, their measured weight was no less memorable. The *Henry Mountains* made Gilbert's reputation in his own day; *Lake Bonneville* has preserved it to the present.

Time's Ratio: The Meaning of Geologic History

In reviewing the *Henry Mountains*, William Morris Davis commented:

> The absence of the important physiographic factor, time, from Gilbert's reports is . . . perplexing. He must have known perfectly well that the existing conditions of drainage systems as well as the existing forms of the land surface are the product of erosional processes acting upon structural masses through

longer or shorter periods of time; yet his account of streams
and of land forms is much more concerned with their existing
status than with their evolutionary development from an earlier or initial status into their present status.

The passage brackets Davis' thoughts on landforms quite succinctly. He is incredulous that, in Gilbert's works, "it is only
by reading between the lines that the idea of systematic change
with the passage of time is to be gathered, and even then but
incompletely."[57]

What characterized Gilbert's interpretation of the Henrys applied equally to *Lake Bonneville*, for what attracted him to Bonneville was not its history but the immediacy of its features. "To the
geologist, accustomed to speak familiarly of millions of years," he
wrote King,

> it was the veriest yesterday when all these things were
> wrought; nor can anyone who stands on the quartzite shingles
> of one of the old beaches, and contemplates the rounded peb-
> bles, gleaming with the self-same polish they received when
> the surf laid over them, fail to be impressed by the freshness of
> the record.[58]

It was as though the lake features were only a frozen moment of
time. If Dutton's monographs were romantic history, then Gilbert's
were slice-of-life realism.

This sense of immediacy held for all the features GK analyzed
at Lake Bonneville—shore forms, volcanism, faulting, and isostatic
rebound. Nothing he looked at had evolved through the stages of a
life cycle; rather than displaying an irreversible history, the geologic
processes were rhythmic, and he had happened to arrive at an instant when the plexus of rhythms made them especially visible. For
Gilbert the dictum that the present was the key to the past had a
meaning peculiar for his time. He was attracted to dynamical rather
than historical geology, uninterested in present events as continuous with a past and a future—instead, he studied the mechanics of
their existing behavior. What impressed him about the gleaming
pebbles of the Bonneville bars was not their vast age but their evident absence of age. It is the same bias shown in his preference for
the timeless forms of mathematics over the time-designating forms
of fossils. He studied the bars as though the waves were still wash-
ing over them and they were dissolving and reforming before his
eyes.

There are exceptions to this observation. Gilbert admitted for both streams and beaches that, after an equilibrium was established, there could be a tendency to degrade and simplify the profile. But it is interesting that neither concept was his own. The degradation of valleys by streams was common knowledge, and that of beaches he borrowed from von Richthofen. More congenial to his own frame of mind was the belief that the tendency toward uniformity (degradation) was resisted by a tendency toward diversity. The resulting landscape was an equilibrium between these antagonistic tendencies. If the landscape were homogeneous in structure, then uniformity dominated; if not, then, depending on the rigor of the climate, diversity dominated. These competing forces were arranged as ratios—his preferred mode of mathematical expression.

It was in his vision of history that Gilbert most differed from his contemporaries, from both the geologists like Dutton and Dana and the geophysicists like Becker. Coping with the onslaught of new information, the earth sciences had taken two directions. On the one hand, evolution—directed by the fossil record—furnished one organizing schema for geology; on the other, geophysics—guided by the newly discovered second law of thermodynamics—provided another. The paleontological schema resulted from the temporalization of the great Chain of Being. Paleontology was the key to the organization of earth history, and stratigraphy, the first geologic science to be arranged by correlation to the fossil record, was the key to the earth sciences. Much of the remaining development of geology involved a search for fossil equivalents that could be organized on a similar pattern, much as physics expanded by hybridizing new subjects with mechanics. This steady assimilation of geologic topics into the realm of *historical* geology is probably the ruling theme to the history of the earth sciences of the period; the assumption of historical causality was, after all, the foundation for geology's claim to be a legitimate science. Nearly every subfield, at least in the United States, followed the example of stratigraphy. Each searched for new kinds of earth fossils, such as landforms or states of rivers, which could be organized along an evolutionary progression.

Geophysics, meanwhile, sought out new forms of entropy clocks, such as heat loss or tidal friction, by which to develop a similarly progressive history. Although geophysics and geology might quarrel, the contest was more over technique and metaphor than over purpose. Geophysics might work within the confines of a contractional hypothesis, geology within the context of evolution, but the ambition of both was ultimately historical. Both were staunchly committed to historicism, whether as something continually gained or continually lost, to the belief that earth history revealed the linear progression of what Arthur Eddington has termed time's arrow.

The landscape of time was as much a terra nova of the second age of discovery as the new continents and oceans it charted. As mentioned above, GK's perception of geologic time, however, led him to solve the historical riddles posed by the earth in ways quite different from those of his contemporaries. Gilbert as classicist, accountant, engineer, and Newtonian philosopher created unique solutions to the landscape discovered by Gilbert the explorer, solutions manifest in his metaphors, prose style, methodology, and theoretical constructions. Especially with regard to assumptions about the nature of history, and with such metaphors as the thermodynamic and life cycles which underwrote the common geologic thought of the day, he had nothing to do. As discussed above, he distrusted paleontology and never practiced stratigraphy as a method of describing history. Even his geophysics harked back to an earlier era in which time was more or less reversible. There was no necessity for the world described by the older Newtonian mechanics to proceed in a particular direction; one could, for example, reverse the algebraic signs of the equations and have the planets move in an opposite order. And it is this perspective which is implied in Gilbert's geologic monographs. Almost ignoring the debate about the age of the earth, he preferred states of equilibrium to stages of evolution. In place of thermodynamic or life cycles, earth history proceeded as a series of discontinuous systems, each bracketed by its own state of equilibrium. Time was not an arrow but a ratio—the relative proportion of history occupied by such systems. It was not a process but a quantity such as mass. Causality was not historical. Instead of entropy clocks, Gilbert sought diligently for such rhythmic chronometers as astronomical cycles or the frictional rhythms arising from the continued action of force and resistance along some boundary. His appreciation for the concept of equilibrium thus supplied him with both a macrocosmic and a microcosmic order: it regulated the basic geologic processes, and it carved up the geologic record into suitable units of analysis. And, not least of all, it dismayed and confused his admirers and memoirists.

Gilbert's organization of space and time revolved around two concepts, rhythms and ripples. The agents of geologic work, temporally organized as rhythms, operated in cycles of varying periodicity. Spatially they manifested themselves as ripples, whose scale could vary from the tiny dunes festooning a sandbar to the domes contouring the High Sierras. The physical explanations for these patterns could be found from mechanical analogies relating force and resistance. Gilbert proposed several candidates for rhythmic timepieces, his favorite being the precession of the earth: the precession could systematically alter climate, and the resulting cycles might be locally manifested as cycles of erosion and deposition. In the 1890s he actually experimented with correlating repeating series of strata in Colorado to such a cycle. Similarly, the earth's surface consisted of undulations—wavelike forms whose scale he expanded to encompass the whole globe in the shape of continents and ocean basins. In fact, he employed the concept of crustal undulations to argue against the reigning belief in the growth of continents and proposed instead that continents only represented oscillations of the crust fluctuating about an ideal mathematical form, the geodetic sphere.

Natural rhythms came from two sources. The basic rhythms were astrophysical and geophysical, giving rise to climatic and orogenic episodes. The second source was friction—the action of force and resistance set up frictional rhythms between the mediums involved, a pulsing action across the earth's surface. One could find such frictional oscillations in the motion of earthquakes, rivers, and glaciers and, by analogy, in nearly all the dynamic processes that sculptured the earth's surface. This mechanical action of force and resistance produced a variety of equilibrium forms—such as the graded stream at the Henrys—and every major monograph by Gilbert pivots on just such a phenomenon. Landforms existed in a state of dynamic equilibrium, their history was rhythmic, and their shape was rippled or wavelike.

GK displayed this philosophical bias in a number of short essays as well. In 1889 he wrote a graceful, popular history of the Niagara River. But his real subject was geologic time. Since the river was "laboring at a task," its history was circumscribed by that task. The falls began when the river cut into the plateau separating Lake Erie from Lake Ontario; when it finally completed its traverse of that plateau, it would cease. Thus, unlike Dutton and Powell, Gilbert did not consider rivers as permanent features of a landscape. Enticed by having the historical boundaries to Niagara so apparently explicit, he attacked the problem of using the rate of recession of the falls as a chronometer—a method he had explored at Cohoes.

Since the age of the falls was a function of its potential work, Gilbert began with an analysis of how the falls operated. "There can be no question that the cataract is an efficient engine," he wrote, "but what kind of engine is it? What is the principle on which it works?" He determined that Niagara functioned like a pothole turned on its side. Were this process constant, the rate of recession would involve only a simple computation. But, in contrast to most treatments of Niagara, Gilbert demonstrated just how profoundly discontinuous this process was. Recession varied with changing volumes in the lakes, with changing rock strata comprising the falls, with isostatic rebound, with shifting outlets to the glacial lakes, and so forth. His narrative ended with a nonconclusion—sixteen questions or minimum variables whose values were necessary before the age of Niagara could be settled. Only by studying the circumstances which created the changing plexus of processes could the age of the falls be determined.[59]

In 1883, Gilbert had used similar reasoning in a devastating review of J. D. Whitney's *Climatic Changes of Later Geological Times*. Where the Niagara essay attacked a geological measure of progressive time, the Whitney review had demolished a geophysical measure. Arguing from proposals by Kelvin and Helmholtz that the sun must be progressively dissipating its energy just as the earth was, Whitney had proposed a uniform degeneration of terrestrial climate and had tried to show how this entropy clock could date various geologic events. Gilbert easily blasted this thesis with some quantitative analysis and, through the use of the graded stream concept, demolished Whitney's most formidable geological evidence—relic stream terraces. He concluded: "If a rise of temperature is not favorable to glaciation, if a fall of temperature does not make deserts drier, and if river-terraces are not indicative of waning precipitation, it might seem that our author's theory is badly off . . ." He politely suggested, however, that the "case is not hopeless."[60]

Gilbert's own vision of geologic processes was made explicit in another popular essay, written in 1888, titled "Changes in the Level of the Great Lakes." This research, nearly a paraphrase of his earlier work for Powell on the Great Salt Lake, provided a popular forum for the theories of shoreline processes he had developed at Bonneville and had special appeal in that it blended theoretical and economic problems. The formula was as embarrassingly familiar as it was simple: there were forces tending to create lakes and forces tending to obliterate them. The two were "so nearly balanced," as he conceived them, "that the scale is thrown to one side or the other by the accident of climate." In a summary passage that might equally apply to *Lake Bonneville*, he expressed a philosophical creed as fully as a matrix of physical facts:

> And so these lakes of ours, that seem to ordinary observation as enduring as the earth and yet as fickle as the weather, are to

the trained imagination of science both ephemeral and constant. The geologist looks backward to the time when they were not, and forward to the time when they will no longer be; talks of their birth, growth, decline, and death, and, comparing their span of life with the earth's, declares them evanescent. The physical geographer, analyzing the motions of the water, refers them to the attractions of celestial bodies, the pressure of air, the friction of winds, the varying dryness of the atmosphere, and varying rain, and assigning each fluctuation to its appropriate cause, lays bare a fundamental constancy to which the navigator and statesman may safely pin their faith.[61]

It was to this same "fundamental constancy" that Gilbert as a physical geographer, or dynamic geologist, safely pinned his own faith in nature.

As the most uncontroversial of men, GK directly attacked the rancorous question of the age of the earth only once. This he did in a presidential address titled "Rhythms and Geologic Time" which he delivered to the American Association for the Advancement of Science in 1900. He had not solved the problem, he assured his audience, and had no new estimates to offer. Although "the problem of our earth's age seems to bear repeated solution without loss of vigor or prestige," he modestly protested that his was a "preliminary question of the available methods of approach." He distinguished between two modes: first, those which follow "the action of some continuous process" and, second, "those which depend on the recognition of rhythms." In the first group he cast both the evolutionary geologists and the thermodynamic physicists. He did not denigrate their efforts but, suggesting that approaches of a different sort might prove useful, he urged geologists to search for natural rhythms by which to correlate geologic events. Given some direct correlations, a general time scale could be constructed by applying ratios from that fixed point. In particular, he recommended the "precessional motion," a planetary cycle of nearly 26,000 years, which "pulses steadily on through the ages, like the swing of a frictionless pendulum." This was valuable precisely because it did not systematically degenerate through time; on the contrary, it was in fact a perpetual motion machine.[62]

But, more than in formal addresses or essays, Gilbert's vision of geologic time was most manifest in the way he practiced geology. On the High Plains in the 1890s, there occurred an incident which perfectly symbolizes his personal methodology. It was past sunset, and one member of the party—a novitiate geologist whom Gilbert was training—had not returned to camp. There was no wood for a signal fire, so a search had to be organized. Orienting himself by the stars, Gilbert walked out a fixed number of paces. Then he turned sixty degrees and paced off the same number of steps. Again he turned sixty degrees. He then easily returned to camp, having successfully inscribed an equilateral triangle.[63]

When many explorers went west, they found themselves awash amid a cascade of new information, exotic landforms, and natural marvels; there seemed to be no familiar landmarks to direct them. They eagerly named new species, created new terminologies, generated new sciences to account for that new information. The vast landscapes, moreover, were trivial in comparison with that still larger terra nova of earth history that was being unveiled. Yet what Gilbert did as a geologist was exactly what he did that night on the plains: he did not see new phenomena so much as analogues of older sciences. Earth history was not a new world of huge proportions that required new principles of temporal organization—what seemed to be a featureless landscape could be navigated with a sound sense of mechanics and geometry, guided by the "frictionless pendulum" of the stars.

"A little daft on the subject of the moon"

Gilbert wrote a surprisingly large number of essays during his administrative tenure with the Survey. Most, however, were mathematical-physical riddles. Two others concerned methodology. Most were products of the library, not of the field. Among the most interesting articles were one entitled "The Chemical Equivalence of Crystalline and Sedimentary Rocks," an experiment with probability theory applied to tornado prediction, an analysis of how birds can soar without violating the conservation of energy, a review of isostasy, a new theory of ripple-mark formation based on an analogy to particles on a vibrating plate, and an investigation into whether the Coriolis force could deflect streams into preferentially eroding one bank (it could). The reductionism inherent in these approaches was also applied to the biological world. As a physical geologist, Gilbert was not unimpressed with the work of "organic processes." On the contrary, he remarked, "based on the resulting deposits, the work of life must be ranked as more important than that of winds, waves, and ice." That is, life processes were an engine like rivers and glaciers. Finally, he spelled out what he meant by physical and dynamical geology—"in the narrower sense . . . the study of the processes of change by which the forms of the surface are created and remodelled." It was the function of the earth sciences, he concluded, to learn *why* these natural phenomena are as they appear, *how* they are produced, and what *laws* govern them. It seeks to understand the relations of mutual dependence which bind them together, as causes and effects, into a vast system, into one great individual mechanism, which is the terrestrial globe itself, with all it contains.[64]

It was in this more philosophical mood, moreover, that Gilbert began to define his scientific methodology. In 1885 he delivered his celebrated paper to the American Society of Naturalists, "On the Inculcation of Scientific Method by Example." He followed it the next year with an endeavor "to illustrate the general process by a description of certain special methods of its application." Analyzing a paper in which William Morris Davis had presented various graphic methods in the study of thunderstorms, he carefully elaborated upon the contrast between "graphic" and "algebraic" processing of data: the graph or map had greater visual appeal, the equation more precision. He demonstrated that each method shared presuppositions on the part of the researcher, and he lucidly showed how each style could be translated into the other—and what the advantages were.

In short, the fields Gilbert had staked out under the Wheeler Survey still remained to be mined. But, lacking time and capital, he worked placers instead of attacking the mother lode. His articles and addresses hinted at major works from the past and pointed to more in the future, but in themselves they did little more than keep his hand in geology. His clever study of ripple marks, for example, stands intermediately between his investigations at the Henry Mountains and his flume experiments; his analysis of earthquake mechanics occupies middle ground between his major report for Wheeler, his examination of the San Francisco earthquake, and his posthumously published Basin Range study. The impact of the Coriolis force on streamflow belongs with his general effort to correlate geological phenomena with astrophysical motions, thus standing between the climatic cycles analyzed in *Lake Bonneville* and his 1900 address on the nature of geologic time. But the calm lamentation that closed his essay on ripple marks was a refrain that terminated nearly every study of interest during the years of his administrative tenure. "Having found myself for some years unable to pursue the subject," he remarked with quiet resignation, "I despair of commanding the necessary time and facilities, and avail myself of this opportunity to communicate my observations to the scientific public, in the hope that they may assist in the elucidation of the subject by another."[65]

Hunting a star. The crater at Coon Butte, from Gilbert's November 1891 field notebook. Doodlings like this were rare, a sure sign that the trip was more leisurely and contemplative than most of his expeditions in the 1890s.

In the end GK would elucidate these subjects himself, but only after another decade and a leap across the continent to California, where the glaciated magnificence of the Sierra Nevada, the natural disaster of the San Francisco earthquake, and the landscape wreckage induced by hydraulic mining allowed him to synthesize the geology of East and West. In the meantime he discovered a surrogate. As the Survey reeled under congressional blows, Gilbert quietly worked away on a large and exciting puzzle: the origin of the moon. He wrote a friend, "I am going to hunt a star."[66]

Gilbert's study of the moon, begun in 1891 with a visit to Coon Butte in Arizona, became a serious project during the period of the Survey's congressional trials with his evening visits to the Naval Observatory in Washington. The subject was not so estranged from his usual line of thought as may first appear. Gilbert held a lifelong fascination with astronomy, which was included in the natural phi-

losophy curriculum at Rochester. Like mechanics, another love, astronomy had early attained a mathematical formalism that natural
history lacked. Similarly, his desire to correlate geological processes
and geological time to planetary cycles demonstrates a fundamental
attachment to astronomical topics, which were a common fund of
conversation with him around the western campfires of his exploring days. Even in the Grand Canyon, his sighting of Venus and his
solar observations had preoccupied him more than the majestic scenery and harried adventure.

This dates his science again as closely derivative from Newton;
the moon, of course, had been a major concern in the *Principia*.
What Newton achieved by establishing an analogy between the motion of the moon and that of a falling apple was different only in degree from what Gilbert attempted in the 1890s by comparing the
topographic properties of the mountains and craters of the moon

and earth. It is appropriate that he first announced his moonlet theory in an 1892 address as retiring president of the Philosophical Society, for the interface between physics and geology had often informed two decades of lively debate for the society.

In truth, as GK himself confessed, he was "a little daft on the subject of the moon."[67] Perhaps his failure to satisfactorily resolve the origin of the crater at Coon Butte haunted his imagination. It is as though, since his investigations reluctantly forced him to abandon meteoric impact as a cause for the terrestrial crater, he projected it onto lunar craters. In any event, as with the laccolith, insight came quickly, and most of his research followed afterward in an attempt to evaluate his idea. It is not surprising that Gilbert rarely spoke of the inductive method or that he formulated his own conception of scientific method upon a series of tested analogies rather than upon the progressive distillation of data by induction. Once familiar with a field, he began with insights which were subsequently tested rather than with experiments which culminated in theory.

His investigation at Coon Butte led him to examine the properties of craters. In particular, "suspecting that it is the scar produced by the collision of the earth with a small star," he had to distinguish the forms of impact craters from those of volcanic craters. This was especially important since Coon Butte lay on the fringe of the expanse of volcanism epitomized by the monumental San Francisco Peaks. The Wheeler Survey had thoroughly familiarized him with the typical volcanic forms of the region, and he recognized that Coon Butte did not conform to the patterns. Unfortunately, he could discover no evidence of a "buried star," so he reluctantly compromised his final explanation: molten lava, perhaps the conduit of an incipient volcano, contacted subsurface waters, creating steam, explosions, and a crater.

Gilbert reinvigorated this field study in 1892 with evenings at the Naval Observatory and experiments in producing impact craters. His astronomical "observations were practically limited to two lunations in August, September, and October 1892, a period affording 18 nights available for work." "Clouds and congressmen," he remarked, were "about equally obstructive." Usually he brought a stenographer along to take notes. Meanwhile, he also conducted experiments: varying angles, materials, and velocities, he dropped, threw, and shot pellets of mud, clay, and lead into dishes containing similar substances in an effort to replicate the impact craters he believed he saw on the moon's face. In December 1892 he delivered his conclusions.[68]

The San Francisco volcanic field, Arizona, 1891. Coon Butte was certainly an explosion crater. That it was located on the fringe of this large volcanic field lent credence to Gilbert's hypothesis. *Courtesy of the USGS Photographic Library.*

"The Moon's Face: A Study of the Origin of Its Features," like his resolution to the Coon Butte question, rejected either volcanism or meteoric impact alone as a primary process. Instead Gilbert proposed an alternative process which subsumed their effects. To accomplish this, he had to subordinate the mere surface features of the moon to the larger process of its origin. This resulted in his moonlet theory. As he pleaded in his address, "The problem is largely a problem of the interpretation of form, and is therefore not inappropriate to one who has given much thought to the origin of the forms of terrestrial topography." He brought to the moon the same ingenuity and patience that were a hallmark in his studies of the earth.[69]

By comparing the forms of lunar craters with those of terrestrial volcanoes and by introducing some clever quantitative measures for further contrasts, Gilbert demolished the reigning assumption that the moon's surface resulted from volcanic activity. In similar fashion, he dismissed the alternative hypotheses, including that which

Crater experiments. Gilbert was interested in the fusion of colliding bodies as well as their angles of impact, comparing these experimental craters to those on the moon and, to his frustration, to the crater at Coon Butte. He performed a number of these experiments in a hotel room when he gave a lecture series at Columbia University, referring to them as his knitting. *Courtesy of the USGS Photographic Library.*

argued that the moon's sculpture had resulted from simple meteoric impact. Yet he did not entirely scrap the impact theory—instead he set out to utterly reconstitute it.

His arguments swung on three hinges: first, the homology of forms between lunar craters and those produced by explosive impact; second, the observation, which he never tired of reiterating, that "all solids are in fact both rigid and plastic"; and, third, an ingenious mathematical analysis of the angle of incidence of "meteors" and the subsequent ellipticity of impact craters. On all counts, the analysis bears remarkable resemblance to his analysis of the laccolith process at the Henry Mountains and to isostatic rebound at Lake Bonneville. His prose shimmers with the same excitement, and GK himself conceded that in the course of the paper he had surrendered his "judicial" pose for an "advocatory" one.

Summing up these observations, Gilbert outlined his theory. Initially the earth was surrounded by a ring of debris analogous to

Saturn's. Gradually this debris began to coalesce into asteroidal moonlets, and these merged by impact to form the moon. By assuming a single plane of orbit for the moonlets, he could account for the production of circular rather than elliptical craters: the moonlets had struck the moon perpendicular to its surface rather than obliquely. To discover this "general law" relating angle of impact and resulting crater shape, he had undertaken the experiments mentioned above, systematically varying angle of incidence, velocity of impact, and softness of material. By allowing for a gradation between plastic and rigid behavior in moon rock, he was able to account for the apparent anomaly between the size of the crater and the amount of material thrown from it. This was reinforced by a careful analysis of the fusion created during the impact. Finally, by allowing the successive impacts to dislodge the axis of rotation, Gilbert could account for the equitable distribution of craters across the moon—that is, the symmetrical configuration of the moon which impact acting only in the equatorial plane of orbit would not produce. In all, his theory was a thoroughly typical effort to encase the explanation of landforms within a matrix of fundamental physical laws.

Gilbert's argument was finely wrought, yet as vigorously fused in its coalescence of observation and explanation as were the hypothetical moonlets. It mixed new observations of lunar features (e.g., Gilbert was the first to report many of the furrows and rills which cluster about the craters) with new analytical techniques. To these must be added a final insight which could only have come from one who had at last polished off a monograph on Pleistocene Lake Bonneville: in depicting the massive eruptions of plastic, solid, and liquid debris from the scene of impact, Gilbert concluded that "sculpturing and veneering have so modified the surface that there is no difficulty in discriminating the craters of later date from those of earlier. The whole topography may be classified as antediluvial and postdiluvial." That is, the discrimination between a "topography of the land" and a "topography of the lake" which he used to unravel the history of Bonneville had its counterpart on the lunar surface if one merely substituted the impact craters' inflow and discharge of solid materials for the exchanges of water in the topographic crater of Bonneville. The salt flat, the outlet, the shoreline, and the mountain all had equivalents:

Thus, by the outrush from the Mare Imbrium were introduced the elements necessary to a broad classification of the lunar surface. A part was buried by liquid matter whose congelation produced smooth plains. Another part was overrun by a flood of solid and pasty matter which sculptured and disguised its former details. The remainder was untouched and probably represents the general condition of the surface previous to the Imbrium event.[70]

Here is the clue to this study. The moonlet theory distilled the techniques of analysis, comparison, and hypothesis which Gilbert had lavished on *Lake Bonneville*. The diagram relating craters to angles of incidence derived from the graphs correlating shoreline to climate; the mechanical analysis of impact explosion and fusion was precisely of the same caliber as his explanation of isostatic rebound; the comparison of crater forms was nothing more than an extension of his comparison of deltas and bars. Thus the lunar maria had a system equivalent to that of the Pleistocene lake. It is as though, frustrated in his efforts to translate the insights of his Lake Bonneville research into the Great Lakes region, GK projected them onto the moon.

"The Moon's Face" is premium Gilbert—a dense miniaturization of *Lake Bonneville*, a telescoped version of his glimpses at the Great Lakes. Neither a flush deposit nor a mother lode, it was instead cautiously processed by intellectual smelting—it marks the beginning of modern theories on the moon's origin; in addition, it stands as a revealing caesura between the monographs of Gilbert's youthful western explorations and the wonderful syntheses of his final years in California. Yet it is for *Lake Bonneville* that Gilbert will be remembered, more than for his moonlet theory. It was a masterly investigation but also a magnificent distraction.

The Great Basin Mess

That mixture of duty and disappointment which characterized Gilbert's scientific career during these years held equally true for his family life. In 1883 his daughter, Bessie, whom he "loved more than anyone else in the world," died of diphtheria. Eventually the disease struck both father and mother as well. Both recovered, but the emotional trauma of Bessie's death stayed with Gilbert, prolonging his convalescence. He retired for a month's rest in the Virginia countryside and, during this time, "fought out" and dismissed any lingering religious sentiment. His frustration in pursuing original research must have compounded his grief by preventing the use of it as compensation or diversion. However, the attention he might have devoted to his family he directed toward science, and he found in the scientific club a partial escape from his painfully decaying family situation.[71]

Even before Bessie's death, Fannie Gilbert had begun to lapse into chronic invalidism. Coal-gas poisoning early in 1883 sickened the entire family, but Fannie recovered more slowly than the rest. The death of her daughter worsened her condition. For a while it stabilized; then she began a slow subsidence of strength which continued until her death in 1899. Gilbert faithfully assumed the administrative duties of the house. His thrift was legendary. Increasingly his wife had to withdraw for visits to convalescent hospitals, and his two sons went to boarding schools, summer camps, and the Nutshell, where Gilbert's brother, Roy, took them in. (Even the Nutshell had become rather hollow. Gilbert's mother had died in 1883, his father two years later.) This required money, as did GK's 1888 journey to Europe and the mortgage payments he steadily made on his house. There is no evidence that his approach was miserly; frugality was imposed as a necessity. And in defiance of that necessity the house frequently served as a depot where neighborhood children received milk and gingersnaps. At the same time, remembering his own hardships in college, Gilbert began to lend money interest-free to needy students at Cornell; he no more objected to spending his money in this way than he objected to mending his own clothes at "800% savings." He absorbed his problems with quiet, even good-humored equanimity.

Gilbert related most to his sons through his work. He employed the younger, Roy, who did not attend college, as a teamster on some of his field excursions, but it was Archibald he remained closest to. It was Arch—named after a beloved colleague and himself an engineering graduate of Cornell—who directly assisted in some of his father's hydraulic experiments in California and who remained a confidant of Gilbert's in later years. Between his work and his wife's debilitation, the boys were away from their Washington home more often than they were present. To his regret, Gilbert became rather alienated from his sons, "except as a source of supplies and a stickler for accountability." But he never lost contact with them, and especially with Arch the bond between father and son endured as had that between Gilbert and his father.

Meanwhile, Gilbert elaborated his scientific ties. By 1894 he was a member of the American Association for the Advancement of Science, the American Society of Naturalists, the National Academy of Sciences (he was elected in 1883), and the Geological Society of America (organized in 1888). The number of scientific societies in Washington proliferated in conjunction with the elaboration and consolidation of the federal scientific establishment. Two decades after the founding of the Philosophical Society of Washington—the parent organization—nearly every scientific discipline had a society in Washington. GK belonged to several, although his favorite, after the Philosophical Society, was the Cosmos Club, founded simultaneously with the Geological Survey on the suggestion of Dutton. Except for the National Academy of Sciences, whose meetings he attended but rarely participated in, Gilbert served as an executive officer in nearly all of these societies at one time or another—often several times. They exploited the identical qualities that made him such a valuable counsel to Powell and the Geological Survey. Though he might regret such an office, he rarely refused its call to duty.[72]

The exchange was not solely one-sided: Gilbert benefited enormously from his memberships. Most of his scientific papers for this period were too brief to publish in comprehensive folios, but they were ideally suited as addresses to the Geological Society of America or the geology section of the American Association for the Advancement of Science. Most were published in the proceedings of these societies; some were reprinted in journals with a wider circulation. This brought Gilbert into closer contact with scientists outside geology and encouraged the sort of hybridization he excelled in. He continually renewed his friendships not only with old comrades from the Philosophical Society of Washingon like Clarence Dutton, Marcus Baker, and William Taylor but with Simon Newcomb, Eugene Hilgard, Charles Schott, Robert Woodward, and Cleveland Abbe as well—astronomers, physicists, meteorologists—the very men he gave special acknowledgment to in the conclusion of his moon study. And unquestionably, apart from intellectual stimulation, there was the sheer pleasure of human company.

Even more effective in this role was the Great Basin Mess. This was launched inconspicuously in 1881, when the members of the defunct Great Basin Division—Willard Johnson, I. C. Russell, W J McGee, and Gilbert—convened in a survey room for brown bag lunches. Occasionally they even sat on bedrolls and packing crates in imitation of a Bonneville field camp. Before long, the mess evolved into an indoor picnic, with a different member packing a lunch for the whole group each week. Soon new members were initiated, until finally the group became so swollen in size and ceremony that a professional caterer and a hired room were required. The Great Basin Mess became a semiofficial institution of the Survey; visiting geologists dined by invitation, and the mess was famed for its good spirits. Yet in many ways the mess only reconfirmed the sweep of

Bonneville in Washington. The Great Basin Mess' farewell luncheon for Powell, 1894. Powell is seated at the head of the table in the foreground; W J McGee is seated at the far left; Gilbert is next to him; and Charles Walcott is seated at the far right. *Courtesy of the USGS Photographic Library.*

Powell's influence; neither Becker, Emmons, nor Barus belonged—members were exclusively from the Powell clique—and the noon ceremony must have rankled as another testimony of what these scientists perceived as systematic hostility.[73]

This Gilbert would have deplored. The function of a society, as he told the Cosmos Club on its twenty-fifth anniversary, was "to bind the scientific men of Washington by a social tie and thus promote the solidarity which is important to their proper work and influence." "The world but imperfectly realizes," he continued, "that its progress in civilization is absolutely dependent on science . . . The influence of our scientific corps—an influence of national and more than national extent—is strong in proportion as it is united, and it suffers from every jealousy and needless antagonism." He toasted this anniversary as the "silver wedding of Science and Culture."[74]

For Gilbert the Great Basin Mess could mean only delight. The ritual helped pacify his longing to be in the field, and it served up the good-fellowship he craved ever more as his family fortunes subsided. It even inspired a mock epic poem which depicted the mountains around Lake Bonneville as prefiguring the mess, feasting on floods and geologic cuisine. Above all, it was science. In the early days of the mess, the director once sent a messenger to summon Gilbert, who at the time was completely absorbed in conversation. When the boy arrived, GK obliviously tossed a clean chicken bone over his shoulder. It struck the messenger, who in turn reported to the director that he had " 'tracted Mr. Gilbert's 'tention." [75]

The mess meant a happy respite from cares and duties, among them the onerous labor of self-control and self-effacement. Gilbert could lose himself in the illusion of science in ways he found difficult at home and impossible in the office. When that messenger appeared, Gilbert was too engrossed in his narrative to notice: he was describing a rare instance of horizontal refraction he had witnessed in the desert.

Powell's resignation in May 1894 terminated an era in Gilbert's personal career no less than in the institutional life of the U.S. Geological Survey. The Great Basin Mess sponsored a farewell luncheon for Powell but, as paleontologist Charles Walcott stepped into the director's office, the Survey began to restructure itself. Gilbert stayed on—though, as Davis has commented, almost in a state of semiretirement.

The damage wrought on Powell's land reforms did not signify the destruction of conservation in American—the Irrigation Survey designated neither the beginning nor the end of conservation, but only a phase of it. The program had been politically premature but instructive; it had been lashed too tightly to Powell's progressively idiosyncratic vision.

In 1891, while attacking the Irrigation Survey, Congress authorized the president to set aside lands from the public domain as "forest reserves." In 1897, following a brief inspection of these reserves by a committee drawn from the National Academy of Sciences, the Forest Reserve Act instructed the U.S. Geological Survey to inventory the forests. The Survey still had the greatest concentration of scientific talent in the government, and Congress drew on its knowledge in the same way that the scientific bureaus which proliferated over the next decade did. It was the continued function of the Survey as a resource conservation agency that sent GK to the Sierra Nevada in 1905, where he brought to a climax the hydraulic and environmental studies he had initiated by measuring the levels of the Great Salt Lake for Powell in the 1870s.

In a similar fashion, the congressional onslaught did not destroy geologic science. The Survey was already rent by decisive internal schisms as well as by its political critics. More significantly, just as it had eclipsed the state surveys by consolidating federal geologic science, so the new universities were eclipsing it as a scientific research institution. It is perhaps symbolic that the University of Chicago appeared as the Survey passed its zenith. Staffed with many former Survey members, directed by Thomas Chrowder Chamberlin, and outfitted with a major publication—the *Journal of Geology*—the geology department at Chicago could begin to rival the U.S. Geological Survey. And Chicago was only the most spectacular example. Compound its challenge with that from other universities and from private research institutes, like the Carnegie. As scientific education expanded, so did the number of geologists and the abundance of scientific information. By 1894, the Geological Survey was too small to contain it all.

Yet Gilbert remained with the Survey. It was a period of retrenchment, but he refused invitations to join university faculties. He had had definite offers from Brown in 1872 and Cornell in 1892, which he had declined, probably out of deference to Powell in the case of Cornell, but in 1894 he could almost certainly have revived those offers. Meanwhile, on several occasions he presented lectures on geology at such institutions as Johns Hopkins, Cornell, Columbia, and Vassar. By remaining he strengthened the credentials of the Survey but denied himself the opportunity to establish a school of geologic thought, such as emerged under Davis at Harvard, Chamberlin at Chicago, and Dana at Yale. Though he successfully placed articles in the new periodicals, he lacked the house organ enjoyed by Chamberlin with the *Journal of Geology*. He might have occupied an analogous position within the context of the Survey but abdicated such a possibility by asking to return to research. He remained a genial scientific loner.

Perhaps it was for the best. Gilbert had little more inclination at age fifty-one to teach than he had had at age nineteen. Temperamentally, he was a researcher, not a pedagogue. His mind continued to flash with insight, no duller for its relative preoccupation with administrative routine. His health flourished. Judging from photographs, though he grew more portly around the waist, he remained powerful across the shoulders. "In the ten years of Western mountain work and camp life," he explained to a friend, "I built up my constitution so as to be a very vigorous man. Eight years of diminishing outdoor life and increasing desk work have lowered my tone somewhat and a sickness still more, but I am still insurable and fairly active."[76]

Eventually the Survey gave Gilbert back as much as it had taken. But that would follow only after another decade. Until then his career resembled the workings of old Lake Bonneville after its waters had crested and the lake had drained to the lower Provo shoreline where, regulated in its discharge, it worked at a kind of grade, beveling back a horizon it had already well marked.

5. Grade

. . . the treatment proceeds rather from the point of view that the science is a unit, that its one theme is the history of the earth, *and that the discussions of dynamic geology, physiographic geology, etc., apart from their historical bearing lose much of their significance and interest.*
　　　　　　　　—Thomas Chrowder Chamberlin and R. D. Salisbury

We saw what we had eyes to see. Our point of view was the measure of our perception and appreciation.
　　　　　　　　　　　　　　　　—Grove Karl Gilbert

A "buried star"

"The errand is a peculiar one," Gilbert had written a friend in 1891, as he set out "to hunt a star." But at Coon Butte in Arizona he had failed to discover one. After the moonlet theory and some speculations on method which followed from the investigation of the Coon Butte crater, Gilbert found no star to substitute. He lamented that there was "no probability that I shall ever complete the Pleistocene studies I began in the Erie and Ontario basins." He had allied his career with that of the U.S. Geological Survey, and the future suggested that the two would subside together.[1]

With the retrenchment of the Survey, however, Powell proposed to send him into the field for the 1893 season. Part of the criticism against the Survey's irrigation program was that it had ignored artesian wells, especially on the High Plains, so, in one final attempt to rally his forces before being compelled to resign, Powell dispatched the old guard of his corps to those sites. For three seasons Gilbert labored near Pueblo, Colorado, in what was literally an inversion of his talents. Instead of the structural variety of the Basin Range, he faced the monotonous High Plains. In place of great surface lakes and rivers, he pondered the subsurface fluctuations and currents of aquifers. Instead of releasing his exceptional powers for original investigation, he edited existing theories into a diluted form suitable for popular consumption. His new errand, like the old one, was indeed peculiar, but he found no more evidence of his hoped-for star submerged beneath the plains than he had in the Coon Butte crater.

There were plenty of irritants in the field. For one thing, the mosquitoes were abundant and voracious. For another, a local teamster nearly triggered Gilbert's temper—a mighty achievement—by trying to raise a contracted wage at a time when he thought he had Gilbert at a disadvantage. But GK was as strict in his accounts as ever. When his son Roy joined him, the boy tried to supplement the government rations by shooting rabbits. Gilbert charged the Survey only for those bullets which found their target. And, once again, perhaps the most saddening grievance, the topographic map of the area was so "incomprehensible" that he had to resurvey the terrain.[2]

Nevertheless, from the time he arrived in June 1893 until the winter drove him out at the end of October, he relished the change of work as fully as he thrilled to the distant panorama of the Rockies. For two weeks he was joined by engineer Frederick Newell, who, the following year, incorporated Gilbert's brief work into a major inventory of the water resources in the public domain. Later in the season, Gilbert rendezvoused with geologist Whitman Cross, who was working in the Pike's Peak quadrangle; together they journeyed along the Front Range for a few days. In October, A. H. Thompson visited the camp at Antelope Buttes, Colorado.

Gilbert glowed with the experience of "luxuriating in camp life on the Plains." For him that aesthetic, like his science, concerned the scene's "physical evidence." "Everything but the entomology," he wrote a friend, "is charming." So much so that he could ask rhetorically: "What is happiness?"

"The soul's calm sunshine." True enough, but too abstract and metaphoric; give us something specific. Well, specifically, happiness is sitting under a tent with the walls uplifted, just after a brief shower, when most of the flies have quit lighting on the lobster-red wrists burnt during the morning ride, and gone off to see what the cook is going to do next, and when the thirsty air is rapidly exchanging its heat for the moisture left by the shower. It is rising at 4:30, while Jupiter is still palely visible but there is no longer any temptation to hunt for the comet, taking a sponge bath in the open, breakfasting from off a box lid gaudily decked by a painted table cloth, and then sallying

forth on the white horse Frank to study the limits of the alluvial veneering on the base-level mesas, measure the dips of
rows of rusty nodules, sketch problematic buttes, and gather
the houses of Ammonite, Scaphite, and Hamite. It is going to
bed by early candle light in the midst of a grove of Rhus tox,
hunting the double stars near Lyrae and Cygni among the
branches of overhanging cottonwoods, moralizing on the development of character through the trying associations of
camp life, congratulating yourself that you are not a pessimist,
and finally dropping off to sleep.[3]

So he did not protest too loudly when Director Walcott, as part
of a reconciliation program with Congress, stationed him on the
plains for the 1894 and 1895 seasons. In fact, the defunct Irrigation
Survey had become the Division of Hydrography by legislative de-
cree in 1894. This was but a phase of the Survey's participation in a
national inventory of resources—minerals, land, forests, and water.
But under any pretext a tent was preferable to a desk, even at age
fifty-one. With Newell in charge of the new division, Gilbert pursued his researches into all the economic topics he could generate
out of High Plains geology. He diligently searched out building
stone, fireclay, potential shale ores, and crucial sites for artesian
wells, while supervising the geologic map for the district. For his
first two seasons in Colorado he mapped; for the third he made a reconnaissance, a traverse of the Arkansas Valley to Kansas.
    Once he was in the field, Gilbert also helped train novice ge-
ologists for the Survey while he scanned the terrain for familiar
subjects. He uncovered several, yet they were but separate riddles
serving as annotations to his larger works. Only one was original.
Published in 1895 as "The Sedimentary Measurement of Cretaceous
Time," it was an attempt to organize a stratigraphic column on the
plains rhythmically rather than progressively. Taking an idea partly
borrowed from Newberry, Gilbert tried to correlate the sequence of
depositions, a monotonously repeated cycle of limestone, shale, and
sandstone, to the sequence of climates that the precession could
generate. The essay served to underwrite the more formal discussion he shaped in 1900 with "Rhythms and Geologic Time."
    But all was not geology. During the three years that he rode "the
white horse Frank" over the plains, GK packed off dozens of crates of
fossils, archeological curios, and live specimens of fauna to the national office. The whole enterprise was almost a recapitulation of
his years on the Ohio Survey. But even collections of this sort could
be hazardous. "At noon Venus [one of the pack animals] was bit-

ten by a rattlesnake on the chin," he jotted into his field notebook, "swelling about head ensued, but she kept strong and was in use 5 days later."[4] Unpacking could be equally risky for a Gilbert correspondent. Snakes (no rattlers, however), tortoises, a tarantula—all destined for the National Zoo—were shipped, along with appropriate geographic information in cover letters.

In spite of these opportunities for fieldwork, it seemed as though Gilbert was off the Survey payroll as often as he was on it. Repeatedly, for months at a time, he absented himself by taking leave without pay. Some of the time he spent delivering lectures at universities as well as public addresses. Some went to the composition of dozens of articles for Johnson's *Universal Cyclopedia*. Some went to travel for both pleasure and fieldwork. The rest of his leave time he devoted to the lands around the Nutshell. He was determined to solve the puzzle of Niagara, and, since the falls had nothing to do with artesian wells in Colorado, this meant he had to work on his own time. Field examination proceeded slowly but persistently, and unfortunately the resulting analysis was no less fragmentary. Just as the artesian springs of Colorado spun off a series of brief, mostly reportorial papers on a variety of topics, so did the surface cataract at Niagara. What Gilbert accomplished was professional, but the articles were little more than blazed trees on the major paths of his thought.

The survey of western and upstate New York was laborious. Its purpose recalled Gilbert's years at Lake Bonneville, but its style was that of an individual field geologist, not of a leader of well-equipped field parties. Public transportation made rapid reconnaissances possible, but modification of the land by humans, glaciers, and vegetation obscured its detailed structure. "A topographic map," he wrote in exasperation, "is the prime need." Nevertheless, searching for outlet channels, deltas, terraces, and the records of glacial climate preserved in moraines rather than in lake shorelines had a familiar rhythm. Considering his isolation, he made several notable contributions, especially in unraveling the sequence of outlet channels and relic drainage lines of the Great Lakes. More importantly, perhaps, in this decade of worries, he found comfort in the proximity of the Nutshell, where his brother still lived. There, as at home, he usually read in the evenings. In 1895 his tastes included Kenneth Grahame's *The Golden Age*, Benjamin Kidd's *Social Evolution*, William Dean Howells' *A Boy's Town*, Anthony Hope's *The Prisoner of Zenda*, and, lest science seemed ignored, *The Life of Sonya Kovalevsky*—a book about the great female mathematician of the century which GK found "very interesting."[5]

Gilbert at High Plains outcrop, ca. 1894. *Courtesy of the USGS Photographic Library.*

On March 17, 1899, Fannie Gilbert died. Her health had subsided steadily for almost two decades, and, as the end at long last approached, she had retired to Florida to preserve what strength she had left. Gilbert was at her bedside when she died. His grief was not worsened by any suddenness of the event, but the final separation left him somewhat adrift. His old comrade Clarence Dutton came to the rescue. Ordered back to the army in 1890, Dutton had squeezed what pleasure he could out of his "virtual banishment" to San Antonio, Texas. In particular, he had cultivated a friendship with a wealthy banker named George Brackenridge, who was so delighted with Dutton's erudition and social charm that he financed a world tour for the two of them. In 1899 Dutton planned a short journey through Mexico. He invited his old partner, William Holmes, and, after he heard about his predicament, Gilbert as well. The invitation probably resembled that which the urbane Dutton mailed to Holmes. "Shall be delighted to see you once more," he had written, "and recall old times when we were young and beautiful and when the roses bloomed—or rather when the coyotes howled and the cactus spines got into our shins."[6]

Gilbert on horseback, High Plains, 1894. *Courtesy of the National Academy of Sciences Archives.*

It was a welcome reunion—all it lacked was Powell himself. They visited ruins and volcanoes and spent the long hours on the train playing cribbage, far from coyotes and cacti. Holmes sketched, Dutton puffed on his cigars, and Gilbert jotted stray observations into his omnipresent notebooks as the trio satisfied their objectives: "to see the great peak of Orizaba . . . the face of the great plateau where the highland breaks off next to the Gulf and get a glimpse of

Popocatepetl on the way."[7] And, of course, inspect some ruins. After the trip, Gilbert hurried north for an appointment with another, more systematic expedition, this time to coastal Alaska.

The Harriman Expedition was a product of American philanthropy. Where Morgan sank his millions into art, Rockefeller into charitable foundations, and Carnegie into libraries and institutes, E. H. Harriman of the Union Pacific brought a part of his to science in a rather peculiar form: the strange expedition to the Alaskan coast that he outfitted with a steamship and brightened with a constellation of 25 distinguished naturalists and scientists. In all the party contained 126 members. Many of the scientists were, like Gilbert, veterans of the old western explorations—men like William Brewer, William Dall, Henry Gannett, and Robert Ridgway. C. Hart Merriam of the U.S. Biological Survey served as executive secretary, George Bird Grinnell cataloged the big game of the coast, John Muir returned to the scenes of the exploits which had made him and his dog, Stickine, national literary figures. Artists included Frederick Dellenbaugh and Louis Agassiz Fuertes. Emerging woodchucklike from his cabin at Slabsides, John Burroughs—Muir called him a mud turtle—braved the foreign landscape long enough to write a narrative for the expedition. Gilbert assumed charge of glacial studies and served on the executive committee. This was fortunate because "the ship had no business other than to convey the party whithersoever it desired to go." He frequently interceded to direct those desires, beginning with the organizational meeting held en route west "in the smoking room of the car Utopia."[8]

If the Colorado studies returned Gilbert to his days with the Ohio Survey, the Harriman Expedition recalled his service with Wheeler. The rapid, two-month reconnaissance of the Alaskan coast amassed some five thousand photographs and countless natural history specimens. GK fully realized both the limitations and the possibilities inherent in the expedition. Consequently, he abandoned any effort to study the glacial field synthetically and opted for a documentary approach which sought to map and measure as many glaciers as possible. This would furnish future surveys with a fundamental standard by which to determine glacial movement. Nevertheless, he could hardly resist some analytical insights. Years of studying geology from the window of a moving train had prepared him well for the passing glimpses of coastline from the gunwale of the ship, and his experiences on the Wheeler Survey had educated him in the art of deciphering geology from the debris in stream beds and alluvial fans. In Alaska he spent many days, and a considerable number of nights, camped on the glacial equivalent—the outwash plain. His talent as a boatman served him in maneuvering the small craft of landing parties in the fjords as well as it had amid the rapids of the Grand Canyon.

Gilbert enjoyed himself thoroughly. He welcomed the informal evening lectures. He relished the overnight expeditions, and the stupendous scenery, illustrated by lantern slides, became the subject of several enthusiastic addresses to friends and societies after he returned to Washington. Perhaps most satisfying of all, he at last confronted the glacial engine whose records in New York and the Rockies he had puzzled over for decades. In all he surveyed nearly forty glaciers. He finished his report in 1901, and the Smithsonian completed publication of the entire multivolume set in 1910.

On July 30, the expedition returned to Seattle, having tabulated some nine thousand miles, covering almost one and a half times the total mileage of Wheeler's 1871 marathon. Gilbert elected to stay in the Northwest for several months more, aided by a railroad pass generously awarded by Harriman. It was a region his exploring surveys had never entered, so he conducted his own brief reconnaissance of the river systems of the Columbia and the Snake. He was most curious about river terraces, the Cascade Range of the Columbia, and the submerged forests—which both Newberry and Dutton had examined previously. These formed the topics of a few short papers presented to the Philosophical Society. At the same time, he scanned the basalt floods of the Columbia Plateau—another lake of sorts—and attempted an explanation for windblown sand dunes, as he had for fluvial dunes, by the operation of mechanical fluid forces. He had long been exposed to the dunes of ancient sandstones in Utah, Arizona, California, and New York—as with the glacier, he now had the opportunity to study the phenomenon in operation. His last chance had been in less propitious circumstances—in Death Valley most of the sand had lodged in his eyes.

Before he left the West Coast, Gilbert made an important journey to California. He inspected the universities on both sides of San Francisco Bay, Stanford and the University of California. At Berkeley he rubbed elbows with A. C. Lawson and Joseph Le Conte. The liaisons which arose from the Harriman Expedition later decided GK's pattern of residency: when in California, usually he lodged at the faculty club at Berkeley; when in Washington, at the residence of C. Hart Merriam. But in 1899 he was still homeless. He occupied a room in a Washington hotel and otherwise turned to his brother in Rochester. He was at the Nutshell when the new century arrived.

Yet the unanswered questions of the old century persisted. In

1901 Gilbert returned to the Great Basin to continue his investigation of Basin Range structure. The study was long overdue. The explication of the Basin Range had advanced only slightly since his preliminary theories were announced in his 1875 report for Wheeler. At that time Gilbert had little opportunity for a detailed examination, and he had had virtually none since then except as incidental to his study of Lake Bonneville. In the meantime, debate over the proper explanation for the structure had intensified. Gilbert had refrained from correlating his theory with the broader context of earth evolution. But those who followed had no such scruples. King dissented with Gilbert's pattern of crustal blocks bounded by normal faults by translating the Appalachian model to the region. The sequence of mountains and basins thus became huge, if complex, folds. The advantage to this model was that it resulted from compression, or crustal shortening, and thereby harmonized with the model of a contracting globe. Le Conte similarly tried to rationalize the mechanics of the region with the nebular hypothesis; this meant an analogy to a flexed arch which then collapsed into a chaos of blocks. Powell and Dutton meanwhile compromised the issue by absorbing Gilbert's structural model into an evolutionary chronology. What Gilbert built up, they in effect tore down by asserting that the uplift was an event bounded by erosion surfaces. Dutton posited two, thus equating the Basin Range history with the two-cycle history of erosion and orogeny he had fashioned for the Grand Canyon. Finally, in 1901, J. E. Spurr attacked Gilbert's geomorphic evidence and structural interpretation.[9]

The attack was blunt, reminiscent of the baiting A. C. Peale of the Hayden Survey had engaged in during the Henry Mountains and the early Lake Bonneville studies. GK conducted himself in 1901 as he had in 1877. Privately he considered Spurr, like Peale, almost a crackpot; officially, he proceeded with caution and conciliation. Actually Spurr had revived a variation on King's interpretation that the Basin Range recapitulated the structure of the Appalachians—that is, the mountains resulted from horizontal compressions which created broad anticlines whose centers were eroded away. So great, in fact, was this erosion that Spurr considered the mountains a remnant feature rather than a structural one. Samuel Emmons joined the issue by publicly defending Spurr. Reviewing Spurr's paper before the U.S. Geological Survey published it, Gilbert typed a thorough critique—and his rejoinder showed that the problem still had considerable vitality for him. In defending his original interpretation, moreover, he ignored the evolutionary cocoon Dutton and Powell had spun for it. Rather than cycles of erosion and uplift, he was struck by the continuity of the uplift. Just as the Bonneville deltas impressed him by their freshness, so did the scarps that designated the fault planes of the Basin Range.

The issue was fortunately controversial enough that the Survey detailed Gilbert to the Great Basin to prepare a rebuttal. Willard Johnson, his old topographer from Powell Survey and Bonneville days, joined him as he retraced his steps to the Fish Springs and House ranges that Wheeler had marched him past. The pleasure in camp life which he had relearned on the plains here mingled with the delight in resurrecting an old subject which he had buried with extreme reluctance. "Among my interesting finds," he wrote a friend, "are a number of mistakes made by Gilbert, one of the Wheeler geologists, in 1872; but he was substantially on the right track as regards Basin Range structure." So his humor revived, even as he encountered a familiar dilemma: "The only remedy I know for a Utah desert wind is to camp in the lee of an irrigated farm, and that raises the question of preference—mosquitoes or dust." [10]

Geology lost its stuffiness, its maddening abstraction of mere editorship, with the chance to see and even experience geologic processes again. At the House Range Gilbert almost saw a flash flood, but he only got "some scars trying to lift the roof of a cave. I was in the cave for shelter from a rain storm when I heard the roar of a storm-torrent. It was two miles away, but I thought it close and started to run and see the spectacle. Then I met the cave roof and stopped and sat down again. Fortunately the damage was only skin deep." [11] His wry humor insured that the clubbing Spurr tried to give him would inflict no more damage to his ego.

After his studies on the plains, a new lexicon had appeared in Gilbert's writings which continued into the Basin Range problem. William Morris Davis, building on the insights of Powell, Dutton, McGee, and Gilbert himself, had perfected the thematic structure of the Geographical Cycle, or Cycle of Erosion, complete with a full dictionary of terms. The terms connoted a genetic explanation of landforms—that is, they were a systematization and an elaboration of what Powell had begun with his taxonomy of streams. As he had with Powell's terms, so with Davis—Gilbert used them, though only descriptively. For Davis, terms like "young," "mature," and "peneplain" denoted stages of development in the life cycle of a landform; they expressed causes. For Gilbert, however, they related descriptively to geologic structure rather than to geologic time: "young" meant rugged; "peneplain" meant a relatively level surface. His critique of Davis' lexicon was identical to that which he had leveled at Powell's and at taxonomies in general at the 1888 International Geo-

Gilbert's photograph of camp life at Desert Buttes, Tooele County, Utah, 1901. The scene is typical of Gilbert's camps throughout his various Great Basin studies. Only in 1917 did he abandon the traditional mule for a motor-car, and that was to visit the Wasatch fault, not its more inaccessible relatives in the desert. *Courtesy of the USGS Photographic Library.*

logical Congress. Powell's terms, he observed, were not founded on a rational theory of streams. The arbitrary classification of igneous rocks by the congress similarly lacked a rational theory of volcanism at its foundation, and Davis' terms, while embedded in an evolutionary theory of landscape development, failed for the same reason. At the heart of the matter was a critical difference in their understanding of stream behavior. For Gilbert, a graded stream was virtually timeless; for Davis, it marked the point in the evolution of a landscape by fluvial erosion at which a stream ceased to downcut and began lateral migration. Nevertheless, for about a decade, which included his glacial work in Alaska and the Basin Range study in Utah, Gilbert spiced his field notes and publications with Davisian terms. After that point, they virtually disappeared from his literature, as had Powell's before them.

GK had occasion to test the Davisian system after he left Utah in 1901. In mid October he traveled south, where he rendezvoused with Walcott at the Grand Canyon. With the legendary William

Wallace Bass as their guide, they spent several weeks in the gorge, much of the time at Bass Camp, along Shinumo Creek. Here Gilbert debated the pros and cons of Dutton's two-cycle theory of erosion and orogeny for the canyon, which hinged on whether the sandstone terrace Dutton had named the Esplanade had functioned as a temporary baselevel or not. Gilbert decided, as he had three decades earlier with respect to the terraces of the High Plateaus, that the Esplanade was a structural rather than an erosional terrace. After he and Walcott left the canyon, he toured much of Arizona and New Mexico by train. In part this was pure nostalgia—between excursions to the canyon, the Great Basin, and the mountains of the Southwest, he was reliving his exploring adventures with the Wheeler and Powell surveys. In part he was also reexamining the broader structural environment of the Basin Range. As a postscript, he swept through southern California and the Coast Ranges before returning to Washington.

At home again, for several months he labored over his maps and notes. Sometime in 1902, however, the maps disappeared mysteriously. Gilbert claimed later that they had been burned—and his hopes for a scientific comeback turned to ash with them. His field notes remained, but they were as spotty as usual. More than was common with him, he had organized this campaign to collect observations rather than to stimulate new analyses. Over the years he had modified his theoretical interpretation of the Basin Range structure only slightly. What he had required was a series of topographic documents to substantiate that explanation and to justify the validity of his geomorphic criteria for faulting. With his maps gone, Gilbert lost the raison d'être for his field work. With a favorite subject once more denied him by an almost inscrutable perversity of fate, the heart temporarily went out of him as well. The Basin Range faded into dormancy again.

But not before he launched one final gambit. Frustrated by an indirect attack on the question, he opted for a direct assault on the structure of the Basin Range. In 1895, as discussed below, he had urged the Coast and Geodetic Survey to extend its network of gravity readings over the Basin Range province, and in 1902 he submitted a proposal to the Carnegie Institution to sponsor a drilling project which would bore deeply into the earth's crust. Neither request succeeded. His proposal to the Carnegie, however, sparked enough interest that he received a $1,000 grant for a preliminary study. In 1903 he selected a suitable site for the project—the batholith underlying the Georgia piedmont near Lithonia, Georgia. Gilbert inspected the area personally and calculated more precisely the cost estimates for drilling. Thanks to the burgeoning oil industry, the technology of boring was advancing rapidly. The scientific value of the project was unquestionable. There was no more pressing question, he observed, than the determination of the geophysical properties of the planet; in an indirect way, the project might also say something about the origin of the moon. The topic became especially acute after Thomas Chamberlin and F. R. Moulton, already subsidized by Carnegie grants, began serious investigations based on their planetesimal hypothesis. A committee of three—a physicist, a geologist, and a man with practical drilling experience—would supervise the boring of six thousand feet into the crust. But on reconsideration, particularly given its $110,000 price tag, the Carnegie Institution decided against the proposal. As GK conceded, the cost was "so large as to be prohibitory." It would be another thirteen years before he addressed the question again.[12]

At the end of what might have been a promising decade, Gilbert's fieldwork had ranged widely but inconclusively. His family life was finally disintegrating: his wife had died in 1899, his brother at the Nutshell in 1901; his two sons had grown into their own careers. As the decade proceeded, the number of obituaries and memoirs he wrote on old scientific colleagues threatened to exceed his production of original scientific papers. Most of his own observations were absorbed into the monographs of others. For a while he seemed to surrender his own distinctive thought to the language and conceptual architecture of the Davisian Geographical Cycle. Given that temporary capitulation, one might appropriately borrow the Davisian interpretation of a fundamental Gilbert invention, the graded stream, to symbolize his status at this crucial juncture in his career. His youth had passed. The currents of his thought ceased to cut down into new scientific bedrock; rather, they swayed back and forth in a broadening floodplain. Gilbert had reached grade.

The Mean Plain

With the exception of some important statements on geologic method, Gilbert's scientific publications during the 1890s generally reflected the two regions of his fieldwork: the High Plains and the glaciated topography of New York and Alaska. On the High Plains, he studied streams, laccoliths, lakes, isostasy, and a peculiar landform, the tepee butte. In New York and Alaska, he examined the mechanics of glacial erosion and documented evidence of its local variations. Collectively, these studies amounted to observational footnotes to his major monographs. He added little that was theoretical, and when around 1900 he decided to coauthor a textbook, it was designed for a high school audience.

The least interesting product of his years on the plains was his 1896 official report, "The Underground Water of the Arkansas Valley in Eastern Colorado." Like several other publications by him in the 1880s and 1890s, this was written in a popular style, deliberately designed for consumption by the local population in Colorado. Carefully, yet without condescension, he described how wells could be profitably drilled along the Arkansas River. Chamberlin had analyzed the theoretical mechanisms of artesian wells in an earlier annual report of the Survey, so Gilbert limited himself to translating theory into practice. In this he excelled. "In many cases," he wrote, "the resident can be his own geologist, and it is worthwhile to point out how he can help himself to the knowledge he desires."[13] In supplying that knowledge, the article was almost as lucidly composed as it was intellectually barren.

Several features nevertheless merit notice insofar as they intertwine with the larger tapestry of GK's thought. For one thing, he conceived of the underground streams in precisely the terms he used for surface flow; the discharge was a function of two competing variables: force and resistance. For another, he reinterpreted the stratigraphic column in physical terms, so that the aquifer (the Dakota sandstone) was bounded by strata as a stream is by banks. Director Walcott proudly remarked in his annual report that Gilbert's predictions were totally confirmed by subsequent drilling. Gilbert also warned about the dangers of overpumping, and these predictions were also, if less happily, confirmed.

In several short essays, Gilbert further extended his equilibrium model to new landforms. The shape of tepee buttes reflected the differential erodibility of two lithological units, their profile a constant ratio of the two. Lakelets represented an ephemeral form created by the competition of wind and rain; as either one or the other achieved temporary supremacy, the lakelet was filled or scoured. Thus, although his themes remained unchanged, their physical size shrank and, along with it, their intellectual significance. It was as though Gilbert, like a Gulliver of geology, had landed in a Lilliputian landscape. Niagara slowed to the trickle of an underground aqueduct, Bonneville became a blowout dune, the laccolith shrank from its majestic size in the Henrys to the relics of Twin Butte in southeastern Colorado. GK seized upon this occasion to reconfirm his original evaluation of the laccolith—even retaining his own original spelling "laccolite." He used the dimensions of Twin Butte to estimate the amount of overburden that must have rested on it at the time of formation, and he reaffirmed his belief, in contrast to Dana, that great viscosity was not the essential trait of major intrusions. The viscosity of known laccoliths represented a condition of preservation, he felt, not of generation. Less viscous (basic) igneous rocks, like those at Twin Butte, were less abundant because they weathered more rapidly than the viscous (acidic) igneous cores of the Henrys.[14]

His most important work on the plains was auxiliary to his formal assignments for the U.S. Geological Survey. In 1894 T. C. Mendenhall, superintendent of the Coast and Geodetic Survey, invited Gilbert "to make suggestions as to the selection of points of gravity stations." The Coast Survey was interested in properly assessing the amount of gravitational anomalies (as measured by the pendulum) attributable to the circumstances of local geology. Gilbert, for his part, was intrigued by the use of geodetic data to test the relative merits of two current theories of crustal behavior: high rigidity, advocated by most geophysicists, and isostasy, championed by most geologists and geodesists. He examined ten stations between the Rockies and the Appalachians on his journeys between Washington and the plains.[15]

It was an old debate, and Gilbert argued a familiar position. According to the hypothesis of rigidity, all material above the geoid meant an excess of gravity which varied as a function of the thickness and density of the rocks between the altitude of the station and the geoid; according to isostasy, the material above the geoid should be compensated so that there would be no excess of gravity. That is, according to rigidity, crust above the geoid meant additional rock of the same average sort; according to isostasy, more rock meant lighter rocks, less rock meant denser. Given crustal material above the geoid, rigidity therefore posited a higher gravitational anomaly than did isostasy. Gilbert corrected his raw data according to the formulas of the two theories and compared the results.

The measurements, he concluded, "are thus seen to be six times as discordant from the point of view of rigidity as they are from the point of view of isostasy." "Nearly all the local peculiarities of gravity," he observed, "admit of simple and rational explanation on the theory that the continent as a whole is approximately isostatic, and that the interior plain is almost perfectly isostatic. Most of the deviations from the normal arise from excess of matter and are associated with uplift." In sum, Gilbert argued that major provinces were in isostatic equilibrium. Minor regions, both with an excess or with a deficit of material, were supported by rigidity. His interpretation only reconfirmed the compromise that he had steadily advanced from the 1870s onward. After Dutton coined the term "isostasy" in 1889, Gilbert crystallized his position in a famous epigram: "Mountains, mountain ranges, and valleys of magnitude equivalent to mountains, exist generally in virtue of the rigidity of the earth's crust; continents, continental plateaus, and oceanic basins exist in virtue of isostatic equilibrium in a crust heterogeneous as to density."[16]

To prevent his analysis from degenerating into an arcane discussion of little significance to field geologists, Gilbert created a new standard of reference. Rather than the geoids or spheroids of geodesists, he designed the mean plain. He considered the central plains—a region long void of orogenic activity—as being in nearly perfect isostatic balance, and he took the average gravitational force in the region as standard. He then measured deviations in terms of the number of feet of a column of average rock (2.70 mean density) which had to be added to this plain to produce the excess gravity. The unit of this new measure, which he termed "rock-feet," determined the amount of uplifted rock not adjusted isostatically—it was a geologic expression of gravitational anomaly. Here once again was the operation of an engineering imagination, giving visual, concrete expression to abstract concepts.

It was something else as well. Here—in geodetic measurements —was GK's true concept of baselevel. The mean plain is his version of the Davisian peneplain. The true measure of erosion and uplift was found in quantitative variations from gravitational equilibrium, not in the form or appearance of level surfaces. Erosion and uplift were continuous, but so—within limits of rigidity—was compensation. What Gilbert saw on the plains was not the record of erosion cycles but the disturbances of equilibrium. His mean plain could serve to show deviations from the normal as fully as could Davis' peneplain, except that the mean plain measured gravity and the peneplain history. The contours of isostatic anomalies furnished a profile of force (gravity) and resistance (local geology). The landscape of geophysics thus shared the same shapes as the landforms of rivers, beaches, glaciers, and hillslopes. Dynamic equilibrium became a gravitational rather than a degradational condition, and Gilbert used the mean plain as a geophysical rather than a geographical measure.

The point can be exaggerated. Gilbert failed to develop the concept into an explanatory system of anything like the breadth that the peneplain assumed in the Geographical Cycle of Davis. Being impressed more with diversity than with uniformity, he was temperamentally unsuited to concoct universal systems. In 1895, however, upon the completion of this study, he urged the Coast Survey, unsuccessfully, to extend its triangulation network over the Basin Range. "It is a question of great interest," he pleaded, "whether the central part of the Great Basin, a broad district of ancient and modern corrugation, but without net gain or loss from surface action, is in equilibrium." What geomorphology could not measure about earth structure indirectly, he hoped to attack through more direct geophysical techniques, such as drilling or gravity determinations.[17]

Gilbert spent his facts as he did his dollars: every one was accountable, and he usually saved them for major purchases. Nothing was wasted. If closeted in Washington, he studied the texture of the moon; if stranded on the plains, he studied the gravitational geometry of the earth. His glacial work repeated this pattern. He intended *Glaciers and Glaciation*, volume 3 of the Harriman report, to serve two functions: on the one hand, his photos, maps, and measurements sought to give an accurate description of glacial locations as a standard for later surveys; on the other, he explored some mechanical analyses for glacial flow and some of its peculiar topographic forms.

GK's contribution, completed in 1901, ranged over three topics: glacial climate, glacial topography, and glacial motion. Collectively it was a substantial addition to what would be an escalating literature of reconnaissance. "The growth of knowledge of Alaska glaciers," he remarked, "is so rapid, that a summary of existing knowledge would have but transient value." For many of the glaciers, I. C. Russell, John Muir, and H. F. Reid had preceded him; instead of synthesizing the entire field of knowledge, however, Gilbert merely placed his volume next to theirs.[18]

First, he manufactured a plausible "climatic explanation" for the irregular behavior of Alaskan glaciers. He observed that "the combination of a climatic change of a general character with local conditions of varied character, may result in local glacier variations which are not only unequal but opposite." For that general change of climate, he suggested a variation in the temperature of ocean currents rather than of air masses. To make this point, he hypothesized the consequences in the form of a scenario: altering the temperature of the ocean a few degrees caused different responses of either advance or retreat, depending on the local circumstances of the glaciers. In short, Gilbert handled glacial climate with the same formula he employed for evolution, landforms, and gravity: the surface of the earth expressed a competition between uniformity and variety. A large force—climate, gravity, or erosion—dissolved into a mosaic of particularized forms according to the local resistances offered it. This was his answer to the theories of glaciation offered by J. D. Whitney, James Croll, Thomas Chamberlin, and others. A glacial climate, like gravity, might represent a global force, but its effects were equally a function of local conditions. In this way Gilbert's analysis of Pleistocene glaciation complemented his investigation of gravitational anomalies.[19]

In delineating topographic features, Gilbert borrowed the insights of Henry Gannett on the effect of tributary glaciers, then coined the term "hanging valley" to designate their subsequent form. This minor episode, however, may be taken as symbolic: what he had done with the fault scarp and fluvial landforms he repeated for glaciation by introducing geomorphic criteria for identifying past geologic processes. As for the coastal landscape at large, he related "the system of relief . . . to three known base-levels." These base-levels he referred to as peneplains, but he soon made clear the peculiar sense in which he employed the term. The "lower peneplain," he wrote, "included all phases of the topographic cycle, being infantile to adolescent where the rocks are most resistant, adolescent to mature in the greater part of the Alexander Archipelago, and senile where the rocks are weakest." In brief, Gilbert reduced Davis' temporal stages to descriptions of structure. He recognized, however, many landscapes that demanded a less simplistic correlation, and he reverted to his fundamental formula. The force of the glacial system was opposed by structures of varying resistances—where the velocity of the ice was powerful enough, it overcame all resistance; where slight, it failed to budge even weak strata. The landscape reflected the ratio of these competing variables.[20]

When he attempted to describe glacial motion, Gilbert adjusted old hydrodynamic concepts to the new engine of glaciation: the glacier behaved like a stream, except that viscosity rather than momentum was the decisive variable. Glacial work proceeded by processes of both abrasion and plucking. The outcome of glacial activity, as with streams, was to equalize its work or "to reduce the profile of the bed to simple forms." The chief measure of glacial energy was velocity, and "most of the inequalities of velocity are determined by gravity in conjunction with the friction of the ice on the channel and the resistance of ice to internal shear; and the processes are essentially the same with water." Like rivers, the glaciers adjusted to existing topographic structures. That is, velocity (and thereby erosion) increased where there were prominences and decreased where there were cavities. The outcome could be termed a "graded glacier."[21]

Perhaps the most original feature of this exceptionally condensed volume was Gilbert's stress analysis of tidal glaciers. He reconstructed the static equilibrium beneath the glaciers—the gravitational pressure of the ice, the hydrostatic pressure of seawater, and the molecular forces of basal meltwater. With the aid of laboratory experiments suggested by Becker, he demonstrated that the thickness of the basewater varies with the pressures of the glacier and the ocean. Where thick (that is, "of more than capillary size"), the water could be partly sustained by hydrostatic pressure communicated by the ocean; where thin, as it most commonly will be, it is, "in some sense, an elastic spring or cushion interposed between the ice and the rock. It performs its function of transmitting the pressure of the glacier to the rock bed quite independent of the presence of the sea." The upshot was that tidal glaciers could erode beneath sea level and that the existence of a fjord "is not demonstrative of a relatively low base-level at the time of its excavation."[22]

The Harriman report neither initiated nor terminated GK's glacial studies, but it was his single, most comprehensive statement on glaciation. Yet the expedition had been a reconnaissance not unlike the Wheeler Survey, and Gilbert's report resembled those he had written in the early 1870s. Even its subject matter was uncannily similar. Gilbert worked along shorelines amid a backdrop of mountains, studying the outwash plains and moraines of Alaska as he had the deltas, fans, and bars of Utah. He even depicted Muir Glacier as an Alaskan Bonneville—a lake of ice, fed by mountain glaciers and discharging through a constricted outlet. He divided the topography into two major categories: the topography of water and the topography of ice. The discrimination between the two sets of processes thus gave him the sort of fundamental contrast that his mind worked best with. "The adjustment of channel contours to simple curves," he noted, "is brought about in both classes of streams (glacial and fluvial), by the more rapid erosion of projecting angles, but the work of water is concentrated on these through the property of momentum, and the work of ice by viscosity."[23] In the end the Alaskan coast was a landscape in equilibrium between eroding energies and resisting structures. In his studies of New York glaciation, he echoed the same conclusion, disintegrating the continental sweep into a conglomerate of local responses determined by varying resistances.

During the 1890s, Gilbert generally popularized his geologic thoughts. He gave numerous lectures—sometimes single public addresses, as on the subject of Niagara, occasionally whole lecture courses at major universities. At the same time he began writing for Johnson's *Universal Cyclopedia*. In 1901, he organized this material into a textbook. Le Conte's *Elements* still dominated the college field, so, together with Albert Perry Brigham, Gilbert wrote a high school text, *An Introduction to Physical Geography*. Brigham, professor of geology at Colgate, had previously authored a textbook, so Gilbert had some guidance in finding the proper level of complexity. "The order of topics," they wrote in the introduction, "has been adopted after deliberate consideration." It mirrored his own preferences exactly: the earth as a planet inaugurated the book, followed by the agencies of dynamical geology and geomorphology, with the traditional subjects of geography concluding the volume. Since the book was advertised as a text in physical geography, the absence of historical geology and paleontology was not problematical. A "Teacher's Guide" provided a rich bibliography of references, drawn heavily from the documents of federal western surveys. The frontispiece was a photograph of the Grand Canyon.[24]

When Powell died in 1902, the news came as another in a long series of personal losses. An increasing number of GK's professional articles were obituaries; for Powell alone, he wrote four memoirs and edited a collected set of memorials. Since his return to the field in 1893, Gilbert's work had been far from trivial, but it was also far from equal to his great monographs. It was as though he were putting his career in perspective. He had retraveled many of the paths from his exploring days, reinvestigated some of the concepts he had announced as a youth in the West, and redefined that glamorous past in a series of footnotes, both technical and popular, to his major studies. He had tidied up his own legacy to the point where, it might seem, he was preparing to write his own scientific epitaph.

The Inculcation of Scientific Method

In 1896, Gilbert polished off his most philosophical undertaking: an exploration into the nature of scientific research itself. His formal statements on methodology began with an 1883 review of two of Archibald Geikie's geology books. In the course of the review, Gilbert framed his commentary by alluding to the proper scientific attitude (which Geikie had). "The principles which distinguish modern scientific research," he pondered,

> are not easily communicated by precept, and it is by no means certain that they have yet been correctly formulated. However it may be in the future it is certain that in the past they have been imparted, and for the present they must be imparted, from master to pupil by example; and whoever in publishing the result of a scientific inquiry sets forth at the same time the process by which it was attained, contributes doubly to the cause of science.[25]

Gilbert doubled his own contribution in the decade that followed the review. In 1885 he delivered his most significant statement on methodology in a presidential address to the American Society of Naturalists entitled "On the Inculcation of Scientific Method by Example." The following year he repeated the performance in a more practical vein with "Special Processes of Research," published in the *American Journal of Science*. As he lightheartedly interpreted it, the previous address "had not been so clear and convincing as was desirable, and . . . you had delicately given me an opportunity to supplement it."[26] A decade later, in a presidential address to the Geological Society of America, he condensed the two earlier papers into "The Origin of Hypotheses." "The Inculcation of Scientific Method" sprang from his Bonneville studies, "Special Processes of Research" from work like the computation of hypsometry tables, "The Origin of Hypotheses" from the moon study. Their significance ranks exactly with the importance of their source studies.

In addition to being classics in the methodological foundations of geology, "The Inculcation of Scientific Method" and "The Origin of Hypotheses" are accurate descriptions of how Gilbert personally conducted his research. Once again, with his usual undue modesty, he disowned any startling originality. "I shall merely attempt to outline certain familiar principles," he announced, "the common property of scientific men, with such accentuations of light and shade as belong to my individual point of view." While the broad topic was methodology, GK particularly explored the interface between teaching and research. The educator and the investigator shared certain common needs and roles: the educator had both to store minds and train them; the investigator required a mental storehouse of information as well as a trained mind. The great researcher was, by example, a great teacher; the teacher should use that example to communicate scientific methodology. In presenting this process of thought, the researcher became further educated. This was as close as the flunked schoolmaster from Jackson, Michigan, could come to a philosophy of education. And in the distilled prose of the address was the preface to his real textbook—the lifetime of examples provided by his essays and monographs. At the same time, investigation and education were not identical. In particular, the teacher "must not doubt, he must know." Under this compulsion, Gilbert warned, "he naturally and unconsciously acquires an undue confidence in results that have simply arisen from the weighing of probabilities."[27]

Underlying Gilbert's stance was his realization that, just as the knowledge of the world was too large to be encompassed by a single cosmology or principle, so there existed no single method for establishing the scientific relationships evident in the data. On the one hand, there was a problem in even determining the data. "It may be doubted," he wrote in "The Inculcation of Scientific Method," "whether there is such a thing as unscientific observation." One merely observed. A record of untainted fact was thus impossible: the very constitution of the mind and of language precluded it. Scientific observation, moreover, was also selective and concentrated on existing classifications and theories. In this regard psychological bias was not only inevitable but desirable. "We observe and we theorize. But while observation and theory may logically be distinguished, in practice, they must be intimately combined or the best results are not secured." It is through "the observer who is also a theorist, and the theorist who is also an observer," he continued, "that real progress is achieved." The result of all this "creates a tendency, and that tendency gives scientific observation and its record a distinctive character."

On the other hand, the relations between such data were no less tentative. If the isolation of a pure fact was impossible, the determination of accurate relationships, especially causal, between such data was very complex. Gilbert identified two means by which observations could be organized: induction and relation. Induction, however, merely led to superficial relations; it was relational classification which led to logical or rational understanding—obviously the goal of science. At this point he introduced another original insight. Relational classification could be of two forms: it could be linear, that is, "linearly arranged in chains of sequence," or it could be coordinate, that is, "coordinately arranged in natural classes" or "a group of coordinate facts having the same antecedents." GK opted for the coordinate form, which better expressed his understanding of causality. He held that "phenomena are arranged in chains of necessary sequence" but that "antecedent and consequent relations are therefore not merely linear, but constitute a plexus" which pervades nature. A simple cause-effect relationship no more existed than did an unadulterated fact. "I am satisfied," he once affirmed, "that all our results in geology are tainted by the tacit assumption of simplicity that does not exist."[28]

Neither simple induction nor universal generalization could cope with the expanding complexity of information and relationships. The first succumbed to the overabundance of information, the second to the overrich interrelationships among the data. The required compromise was the working hypothesis—a scientific guess, necessary both to discover facts and to test relationships. Research consisted of an unending process of generating and testing such hypotheses, and the process of verification became "a process of elimination, at least until all but one of the hypotheses have been disproved." Gilbert concluded that, "if the investigator is to succeed in the discovery of veritable explanation of phenomena, he must be fertile in the invention of hypotheses and ingenious in the application of tests." The education of the scientific investigator therefore demanded "training to improve the guessing faculty," for the "great investigator is primarily and preeminently the man who is rich in hypotheses." Needless to add, the experienced investigator found "relations of quantity the most satisfactory criteria."[29]

Gilbert the classicist formalized the procedure by which one made such guesses in this way:

> Given a phenomenon, A, whose antecedent we seek. First we ransack the memory for some different phenomenon, B, which has one or more features in common with A, and whose antecedent we know. Then we pass by analogy from the antecedent of B, to the hypothetical antecedent of A, solving the analogic proportion—as B is to A, so is the antecedent of B to the antecedent of A.

The role of experimentation—the creative process of scientific thinking—was to supply and refine the analogies which served as hypotheses. "The consequential relations in nature are infinite in variety," Gilbert recognized, "and he who is acquainted with the largest number has the broadest base for analogic suggestion of hypotheses."[30] When, for example, he compared the topographic forms of the moon to impact craters made by shooting pellets, when he compared the mobile crust under Lake Bonneville to a flexed beam, and when he later tried to extrapolate from laboratory flumes to natural streams by recommending that the flume data might apply to streams which were geometrically similar, Gilbert was establishing just such analogic proportions. It is worth observing that the form of thinking which he proposed for geology bears a striking similarity to the form of his mathematics of ratios, to the neoclassicism of his prose, and to his mechanics of force and resistance.

Given all these variables, Gilbert found it impossible to construct a programmatic method which could produce scientific knowledge. At best one could only imitate the great scientists. In the particular case of geology, he felt it should imitate physics. Hence, science was something one practiced more than one preached. The problem of creating a scientific geology was pragmatic, an exercise in psychological engineering. "Precept unsupported by example," he warned, "cannot be depended on to infuse method." The great works of science, like masterpieces of art, served as examples for emulation.[31]

It is the actual practice of science which informs his papers on method. Gilbert's philosophy was pragmatic rather than metaphysical—he was concerned not with the units of thought so much as with the process of thinking. Unlike the evolutionists, he was not interested in the origins of science. Unlike the philosophers of physics, he did not scrutinize the fundamental concepts and language of science. Except in his personal behavior, he ignored the relation of science to ethics. He did not even discriminate between scientific and nonscientific modes of inquiry except in their tendencies. As though it were a geological landscape, he took scientific thought as a given and showed how it worked; he treated the mechanics of scientific thought exactly as he treated the processes of rivers or erosion—his philosophy of method showed how to use what was known to solve new sorts of problems. Not questioning the origin or the epistemological foundations of physics and geology, he showed how they could be equated by an analogy of proportion. Thus physics and geology were equivalent; both were scientific appositives. Geology was not derivative from physics, nor was physics subsumed within geologic cosmologies. Yet to say all this is not to diminish GK's achievement. In geologic science his methodology essays stand among the standard literature as the *Henry Mountains* does to the study of geologic structures and landforms.

Yet it was hardly his understanding of the scientific fact or even his advocacy of working hypotheses which made his contribution special. His insistence that science was communicated by example and analogy did little to make scientific thinking unique either. What distinguished his thoughts was the belief that science could be communicated only by example. What gave this concept and Gilbert both a special flavor was the unflagging effort to reduce egotism. Bias that was perceptual and linguistic could be publicized but never eliminated; prejudice deriving from self-conceit and glory-seeking, however, could be systematically diminished. This was in fact one of the chief functions of multiple hypotheses: they removed the observer from the system insofar as was possible. The romantic egotism and Promethean contests celebrated by many of his contemporaries were utterly foreign to Gilbert's temperament—he would have found meaningless Emerson's remark that he became a "transparent eyeball," pointless Ahab's maniacal pursuit of a white whale. Similarly, he discouraged their scientific counterparts: the diffusion of universal unities, the pursuit of dramatic discoveries, and the championing of individual insights inflated megalomaniacally into cosmological principles. Rather than projecting himself upon the landscape, he tried to efface himself from it. He genuinely protested efforts to name natural features after him (not always successfully), and even in his prose he habitually spoke in the third person and through indirection, a practice that further punctuated his abstraction from the scene.

Controversy was valuable only when it exposed basic principles and assumptions, not when it polarized around personalities. When his ideas were once challenged at a meeting of the Geological Society of America, Gilbert offered no corrections. When asked why he didn't answer his critic, he explained with a laugh, like the boy tumbled downstairs by a schoolmate, "I didn't want to be too hard on him."[32] Whether in the form of deliberate self-dramatization or as the unexamined biases behind psychological perception, egotism was the opponent of scientific thinking; the whole thrust of Gilbert's methodology was to efface or check it. The chief mechanism was the talent for inventing a plethora of ingenious hypotheses, and this in turn demanded considerable creative skill in making suitable analogies.

William Morris Davis wonderfully described the outcome of this philosophy as it was manifested in GK's own hypotheses. "It was his habit in presenting a conclusion," observed Davis approvingly,

> to expose it as a ball might be placed on the outstretched hand—not gripped as if to prevent its fall, not grasped as if to hurl it at an objector, but poised on the open palm, free to roll off if any breath of disturbing evidence should displace it; yet there it would rest in satisfied stability. Not he, but the facts that he marshaled, clamored for the acceptance of the explanation that he had found for them.[33]

Gilbert conceived of science in the same hypothetical way he did scientific theory: he never held it dogmatically or used it combatively.

What rendered scientific thinking tentative rather than mechanically formal was the operation of the "personal factor," that is, the psychology of the observer. In recognizing this prejudice, Gilbert was not unique. Experimental psychology had developed rapidly in the 1870s; and both Powell and Ward, as he well knew, were insisting on the "psychic" factors of civilization as an intervening variable between humankind and natural law at the same time that he posited the "personal factor" interceding between the investigator and nature. Yet he never, in turn, made his new methodology into a new metaphysics. He even considered C. S. Peirce, who welded speculations like Gilbert's into a philosophy of scientific pragmatism, "a man so metaphysical I should never have tho't of going to him with a practical question."[34]

Nor was the recognition of scientific imagination unique. After Darwin the nature of scientific thought and explanation was reevaluated, and even physicists like John Tyndall gave such popular lectures as "The Use and Limit of the Imagination in Science." What Gilbert achieved was to show how—according to a fundamental Gilbertian formula—the use and limit of imagination interacted upon each other. In 1890 William James published a psychological summa on the scale of *Lake Bonneville*—the famous *Principles of Psychology*. In it he developed the role of unconscious psychological processes in what he termed the "stream of consciousness." In his own way, by emphasizing the psychological bias to scientific thought, Gilbert's methodological studies were an analogue to what James, Peirce, and Ward, among others, were doing. But his understanding of the flow of scientific thought had a peculiar twist. He conceived that the advance of science was opposed by a "natural egotism." He showed that, like rivers in nature, the psychological stream should be graded. The method of working hypotheses was the resulting profile of compromise.

One of the curiosities to "The Origin of Hypotheses" that makes its appearance in this period of Gilbert's life especially appropriate is that he was scientifically wrong in a major piece of research. He had journeyed to Coon Butte in 1891 after preliminary studies by Arthur Foote and, under Gilbert's direction, Willard Johnson. The large crater had meteor fragments scattered around it—the odds of a random association, Gilbert computed, were roughly 800:1. So, when he took leave of absence from the Survey to investigate the crater in 1892, he hoped to find a "buried star." The narrative of Coon Butte followed his successive efforts to test this assumption—by magnetic deflection of a compass needle, quantitative comparisons of the volume of meteor fragments and the size of the hole, comparative crater forms, and so on. In the end, failing to confirm his hypothesis, GK abandoned the idea. That disappointment provided the moral to the essay, for it was precisely such preconceptions that the working hypothesis was intended to work against. Unfortunately, he was wrong. Coon Butte is today known as Meteor Crater.

It is another curiosity to the paper that this fact hardly matters.

As an example of scientific thinking—which was, after all, its purpose—it is unexceptionable, even as style triumphed over substance. So it was with Gilbert's career in the late 1890s: his method and thinking were as sharp as ever, but they lacked substance; they languished for want of a compelling theme. The Harriman Expedition had offered a welcome diversion. The High Plains expeditions had done little more than excite his "western fever." The Basin Range study, after his maps disappeared, was only another disappointment. Even his methodological theories, for all their acute insight into the practical machinery of scientific thought, had let nature deceive him into taking 800:1 odds.

The Text for a University Science: Thomas Chrowder Chamberlin

Those were odds Thomas Chamberlin would have found intolerable. Truth, he insisted, "cannot mislead." And the pursuit of truth led him from his log cabin birth on a terminal moraine in Illinois to the reconstruction of the cosmology of the solar system. His fame as an educator rivaled even his reputation as a scientist, so that by 1910 he and William Morris Davis dominated the American geological scene—the one at the University of Chicago squaring off the foundations for planetary evolution, the other at Harvard planing off the evolution of its physical geography. Together they consolidated the American school of geology. Born the same year as Gilbert, Chamberlin survived ten productive years longer, and their separate careers proceeded in an uncanny counterpoint.[35]

Like Powell, Chamberlin was born on a midwestern farm, the son of an abolitionist and sometime preacher. Like Powell again, he diverted that inheritance into science and education. The natural landscape of Beloit, Wisconsin, where the family moved when Chamberlin was three, powerfully impressed him. Like those other geologists of his generation who came from rural environments, he entered geology through the study of natural history—he remained a "boy of naturalistic bent" throughout his career. His earliest scientific curiosity was stirred by the discovery of fossils. And he never forgot the example of Darwin. Like Powell, he educated himself by teaching, eventually even spending a year in graduate study at the University of Michigan. And, again like Powell, the "wider interests and experiences of the war" forced on him a "wider vision." He too became a scientific captain of industry.[36]

Chamberlin's career advanced rapidly. He graduated from Beloit College in 1866, then served as a high school principal for a couple of years before being employed as a professor of natural science at the Wisconsin State Normal School from 1869 to 1873. In that last year he joined the Wisconsin Geological Survey as an assistant geologist; from 1876 to 1882 he acted as chief geologist. Like many others, such as John Strong Newberry and Edward Orton in Ohio, he blended college teaching with survey administration and geology with natural theology. When not lecturing on natural science at Beloit College—he was on the faculty from 1873 to 1882—or engineering the work of the survey, he often addressed the Second Congregational Church of Beloit on the subject of science and religion. In 1881, still holding his other appointments, he entered the U.S. Geological Survey as geologist in charge of its glacial division; he continued in that post until 1904. During this tenure with the national survey, he advanced his stature as an educator by becoming president of the University of Wisconsin. From 1887 to 1891, like the progressive he was, he modernized the university curriculum. In 1892 he accepted the chairmanship of the geology department at the University of Chicago. With a handpicked faculty and the *Journal of Geology*, which he edited, Chamberlin gave the geology department equivalent intellectual status to the other famous departments of the Chicago school, philosophy and social thought. "The earth sciences are not purely physical sciences," he held. "They concern themselves with life and with mentality, as well as with rocks, oceans, and atmosphere." Granted that scale of vision, it was not difficult for Chamberlin to have geology intersect the thought of John Dewey, George Mead, Charles Horton, Herbert Cooley, and the others.[37]

The path of Chamberlin's scientific career followed the trajectory of one fundamental subject: glaciation. As with Gilbert, this was the landscape he had absorbed as a child. But, where Gilbert repeatedly returned to variants of a few themes which he compared and contrasted, Chamberlin steadily expanded the scope of his single field until it encompassed a cosmology. His early papers mapped the preserved patterns of glacial movement. He was one of the first to recognize that the Ice Age was not a unitary phenomenon, as Agassiz had held, but a series of retreats and advances. His 1888 description of glacial work, "The Ice Scourings of the Great Ice Invasion," summarized the midwestern Pleistocene as *Lake Bonneville* had the Great Basin. Searching for an explanation of multiple glaciation, he studied climatic changes—at first, like Gilbert, adopting Croll's hypothesis and finally, also like Gilbert, rejecting it. Borrowing an idea from Tyndall on the role of carbon dioxide in the atmosphere, he suggested oscillations in the carbon dioxide content which, by varying the amount of heat absorbed, could induce a rhythmic cycle of glaciation which steadily dampened itself.

But evidence mounted for glacial periods earlier than the Pleistocene, and this required a relatively constant atmosphere for long eons of geologic time. The nebular model of cosmology, which postulated the condensation of the earth from a hot gaseous medium, could not supply this condition, nor could it, in the hands of physicists like Kelvin, furnish the magnitude of time required by Darwinian evolution. Chamberlin exorcised both demons by reconstituting the nebular model of Laplace into the planetesimal units which then coalesced by cold fusion to form the earth. The circular basins of the earth's oceans mirrored the lunar maria. By 1906 he and astronomer F. R. Moulton had collaborated to the point that Chamberlin published a full description of the process and included it in his influential three-volume college textbook, *Geology*, which he coauthored with R. D. Salisbury.[38] Still restless, he pushed onward, summarizing his theory of planetary evolution in 1916 with *The Origin of the Earth* and the broader issue of the solar system in 1928 with *The Two Solar Families.*

Again following the lead of Gilbert, Chamberlin abstracted the procedures of his own thought into papers on methodology. Gilbert's "The Inculcation of Scientific Method" and Chamberlin's "Studies for Students" probably constitute the chief contributions of American geology to scientific methodology. Ostensibly the two articles merge neatly in their themes: Chamberlin published a preliminary version in 1890, four years after Gilbert's "Inculcation" and six years before "The Origin of Hypotheses"; "Studies for Students" appeared in 1897. Both men accepted the working hypothesis as the foundation to scientific research, and both urged the multiplication of hypotheses as a check against overattachment to single theories. But there the similarities end. Chamberlin's "naturalistic logic" took him in one direction; Gilbert's more mathematical logic took him in another. Their differences in method express precisely the differences in their sciences.

Chamberlin launched his paper with "two fundamental modes of study": "imitation," by which one accepts the reigning authorities, and "independent thinking," the "endeavor to think for one's self." The choice between these two was a foregone conclusion, for the patriarchal Chamberlin was as large in mind as in body—his thought was as vigorous and athletic as his physique. He had little patience with the "demonstration of a problem of Euclid precisely as laid down," which he interpreted as imitation. Instead he encouraged "the demonstration of the same problem by a method of one's own or in a manner distinctively individual." His scientific methodology belonged squarely with the pragmatic philosophies developed in Dewey's instrumentalism and James' pluralism. Hypotheses were only tools to guide lines of inquiry, and they, like geologists, should learn to occupy a pluralistic universe. He admitted that "it is rash to assume that any method is *the* method, at least that it is the ultimate method." Such was the nature of scientific progress that every theory and method would always be succeeded by a better one.[39]

Thus Chamberlin did something utterly in keeping with his training and temperament: he created a genetic evolutionary scheme to explain his methodology. While declining to forecast the future direction of methods, he cited three phases of their past: the method of the ruling theory, the method of the working hypothesis, and the method of multiple working hypotheses. This procedure was nothing novel: genetic taxonomies, evolutionary chronologies, and universal processes saturated all of Chamberlin's thinking. It is, in fact, his real methodology—as it was for American geology at large. In all his labors to expand the frontiers of evolutionary geology, Chamberlin only did for metaphysics what he did for geophysics. In *The Origin of the Earth*, analyzing the arrangement of planets, he emphasized that "these materialistic records are by no means the only ones, nor always the most important ones; there are *dynamic* relics that are as truly vestiges of processes once in progress as are fossils or strata." In other words, the assemblages of planetary motions are fossil equivalents. Chamberlin proceeded to fashion an evolutionary scheme for these dynamic fossils as Darwin had done for organic fossils, Newberry for stratigraphy, Powell for mountains and streams, and Davis for erosional baselevels.[40]

Yet there was a difference to Chamberlin's evolutionism. Physical chronologies were gradually replacing biological ones. But the change in substance did not cause a similar change in the pattern of geologic history. Even in its ultimate form, radioactive dating, which appeared about the time Chamberlin died, physical dating procedures only furnished an "absolute" chronology, based on an irreversible process, to replace the relative chronologies used previously. In the same vein, Chamberlin proposed another correlation procedure in 1909: diastrophism, or large-scale crustal deformations. In the context of his evolving but closed earth, with its static continents and ocean basins, this concept was perceptive. "Stratigraphy and paleontology thus go hand in hand, each sanctioning the other," he observed. "*Diastrophism lies back of both and furnishes the conditions on which they depend.*" Noting how nicely this interpretation merged with the study of baselevels cultivated by Davisian geomorphology, he confidently concluded: "Diastrophism thus seems to me fundamental both to stratigraphic development and life development. Diastrophic action seems to be the forerunner of both these standard means of correlation." Episodic diastrophism amounted to epicycles on the broader curve of planetary evolution.[41]

Like his colleagues who constructed similar chronologies, Chamberlin saw in this pattern of progress evidence of design and morality. Indeed, he had projected evolutionism about as far as it could physically go, bringing it from the ammonites of Newberry to the genesis of the solar system. All that remained was the evolution of mentality. As Dana and Le Conte had shown that biological evolution proceeded from geological evolution, so Chamberlin argued that a psychological evolution would proceed from his cosmology. Just as he began to substitute more physical criteria for Le Conte's biological ones, so did his text *Geology* supplant Le Conte's *Elements of Geology*. *Geology* closed a century's study of geologic time.

The text had as its theme "*the history of the earth*." Such a vision could conclude only with the evolution of humans themselves, and human history, in turn, could conclude only with the evolution of scientific investigations:

> The forecast of an eon to intellectual and spiritual development comparable in magnitude to the prolonged physical and biotic evolutions lends to the total view of earth-history, past and perspective, eminent moral satisfaction, and the thought that individual contributions to the higher welfare of the race may realize the fullest fruits of their permanent worth by continued influence through scarcely limited ages, gives value to life and inspiration to personal endeavor.

In his science no less than in his politics, Chamberlin was a thorough progressive.[42]

He was no less so in his methodology of science. That quality, at once subtle and distinctive, differentiated him from Gilbert. Where Gilbert took science as a given and studied the dynamics of its operation, Chamberlin insisted that it exhibited an evolutionary history. He would bring the scientific method up to date, modernizing its administration as he had the University of Wisconsin and the geology department at Chicago. When Gilbert examined the nature of scientific explanation, he decided that it proceeded by a process of analogic thinking; Chamberlin saw it as evolving by "independent creations." GK's philosophy of science resembled theories of neoclassical art: science expanded its domain by imitating the great works of earlier scientists. The departments of science were thus linked by analogy—science grew by a series of mirror images, reflecting, in particular, the structure and apparatus of physics into new subjects. The theoretical impulse which led Horatio Greenough to sculpture the American hero George Washington in the garb of Zeus the Cloud-Gatherer led Gilbert to describe a novel mountain, like the laccolith, in the language of Archimedes.

Nothing could be further from the scientific impetus behind Thomas Chrowder Chamberlin. The evolution of science, like the evolution of organisms, could spawn new forms, grander and more comprehensive than those which preceded it. Different sciences were not analogues but newer, more comprehensive stages. Geology was thus the most synthetic of all the sciences—astronomy, for example, was merely its "foreign department"; and physics was valuable chiefly for the new dynamic fossils it contained. When physics conflicted with geology, Chamberlin concluded that there was something wrong with physics—a stance which led him to become a prophet of atomic energy. This view made novelty essential to the workings of science. As with organic evolution, there would be no material for selection or progress without novelty of hypotheses and information. Hence, where Gilbert sought to find close-fitting analogies to established sciences, Chamberlin urged the invention of new explanations and, ultimately, of new sciences and newer methodologies. Gilbert's succession of working hypotheses constituted a set of close approximations, like the terms of a converging series. Chamberlin made that model historical: the approximations came closer to truth with each new effort.

To the methodology of both men there was a moral bias. Each saw that bias entered into scientific investigations because of psychological pressures. But, while Gilbert warned against self-conceit, Chamberlin spoke of intellectual affections. For Gilbert one function of multiple hypotheses was conservative—to check egotism. For Chamberlin the method was more liberal—to encourage self-growth by giving "satisfaction to the moral sense" with "the precision, the completeness, and the impartiality of the investigation." Significantly, Gilbert addressed his papers to investigators, Chamberlin his to students. "As a factor in education," the former university president noted approvingly, "the disciplinary value of the method is of prime importance." In short, for Gilbert science proceeded despite egotism; for Chamberlin it proceeded because of it, in the more acceptable form of individualism.[43]

Gilbert looked on hypotheses as potential solutions to what was ideally a quantitative riddle; Chamberlin saw them as children in the family of the parental researcher. His "naturalistic logic," expressed in such metaphors as the family, combined with a secularized version of his father's abolitionist ministry to alert the researcher that "by his parental relations to all" he "is morally forbidden to fasten his affections unduly upon any one." To Gilbert's mind, a large number of hypotheses served to check and regulate, like the graded bed of a stream, diminishing the scope of each hypothesis. To Chamberlin's, the large family diffused the affections, and "the reaction of one hypothesis upon another tends to amplify the recognized scope of each." The progress of scientific thought for Chamberlin thus paralleled the growth of the individual—competitive, pragmatic, and personal.[44]

Both men recognized the distortions of language and the complexity of causality. "When faithfully followed for a sufficient time," Chamberlin argued, the method of multiple working hypotheses "develops a mode of thought of its own kind which may be designated 'the habit of parallel thought,' or 'of complex thought.'" He added revealingly: "It is contradistinguished from the linear order of thought which is necessarily cultivated in language and mathematics because their modes are linear and successive. The mind appears to become possessed of the power of simultaneous vision from different points of view." Chamberlin lamented that such thought could not be rendered into words, but "the remedy obviously lies in coordinate literary work."[45] Henry James, and the chief authors of twentieth-century fiction, such as William Faulkner and James Joyce, have corrected that defect.

For Gilbert, however, the ideal language was mathematics; the perfect composition, a set of axioms. The problem of perception was not to incorporate the perspective of the parental researcher insofar as possible, not to enlarge the moral sense. Rather than an extended family of hypotheses, he imagined a tangled braid of antecedents and consequents potentially explicable by analogies of proportion. The formal language of mathematics provided a medium to erase biasing metaphors and anthropomorphism. Where Chamberlin presented his hypotheses as though they were candidates for a political office, Gilbert offered his as though they were prospective solutions to a mathematical puzzle.

The same appreciation for language was manifested in both men's careful prose. Gilbert's could be fulsome but was balanced; Chamberlin's had a driving energy that pushed on vigorously to its conclusion. His was a large style; he regularly used big words and, like his ideas, his sentences marched on past the expected pauses, often bolstered by strings of prepositions. Compared to Chamberlin's cosmic vision, Gilbert's writing seemed cautious, even anachronistic.

So did his interpretation of glaciation. Chamberlin saw in the Ice Age a change in global climate and planted that episode in the context of his cosmology. That the continental glaciers had advanced and retreated according to some climatic oscillation did not disturb his vision—those fluctuations, like the advent of glacial eras themselves, were little more than epicycles in the larger evolution of the atmosphere, just as diastrophic events were in the evolution of the crust. Hence, Chamberlin studied the astronomical context that led to a global change in climate, while Gilbert, as he did so often and in so many forms, insisted that a global change in climate was meaningful only in terms of local environments. Chamberlin emphasized the general process which overrode particular expressions, Gilbert the local opposition that resolved the universal case into a mosaic of local cases. In the same way Chamberlin, awed by the universal progress of scientific evolution, saw in it a means to amplify individual integrity and moral purpose. Gilbert, impressed by the stubborn resistance offered evolution by individual egos, devised means to constrain them. His working hypothesis was not an evolutionary stage in the progression toward scientific truth but a dynamic intellectual equilibrium between ideas and facts.

The difference between the two men is also apparent in their administrative careers. Gilbert withdrew from bureaucratic chores as rapidly as he could, putting them in the same category as teaching. Chamberlin, however, accepted that equation enthusiastically: he was quick to seize the throttle of social institutions whose machinery could work for progressive values. By 1902 that difference was acute. While Chamberlin occupied positions of scientific and political power, year by year expanding his just influence, Gilbert continued his solitary visits to Niagara, petitioned vainly for an experimental test on his own intuitive version of planetary cosmology, and receded visibly from positions of administrative significance in the U.S. Geological Survey. Chamberlin, progressively creating a larger society around himself at Chicago, watched the universities compete with the USGS as institutions for geologic investigations. Gilbert, living by himself in a Washington hotel, seemed likely to vanish into solipsism.

It is no surprise that the tribute to Chamberlin from his own generation, and from that second generation which included his students, was impressive and genuine. An entire volume of the *Journal of Geology* in 1929 commemorated his career. He was perhaps the major patriarch for the geology of his era, and his textbooks alone would insure that stature for another generation. Though they were never especially close—Gilbert was too much a member of the Washington office, Chamberlin too associated with its rival organization at Chicago—the two long maintained a cordial friendship and exchanged intermittent correspondence. During the World's Columbian Exposition in 1893 and 1894, GK brought "the boys" west, and they stayed as guests of the Chamberlins. And, in his memorial editorial on Gilbert, Chamberlin generously predicted: "It is doubtful whether the products of any other geologist of our day will escape revision at the hands of future research to a degree equal to the writings of Grove Karl Gilbert."[46]

In a backhanded way that observation framed a critique of his own colossal labors. In ways not true of Gilbert, Chamberlin's thought was an abstraction of his own time. When the era of evolutionary geology faded, so did the impact of his texts and his brave forays into the paleontology of planetary dynamics. New questions would be asked. New instruments and data would appear in exponential quantities. Systems analysis would supplant evolutionism. The acceptance of a chance universe would make model building the acceptable methodology of the earth sciences. Yet, when Chamberlin died at the age of eighty-five, the last gigantic figure from the heroic age of American geology, his impression was powerful enough that Bailey Willis justly prefaced his memoir with this scientific honor roll: "Aristotle, 322 BC; Copernicus, 1543 AD; Galilei, 1642; Newton, 1727; Laplace, 1827; Darwin, 1882; Chamberlin, 1928."[47]

An Elder Statesman

In 1902 few people would have made such extravagant claims on behalf of Grove Karl Gilbert. Nearly sixty, he seemed to have eased into a career as an elder statesman of geology—a man honored for his explorer past, given more to the philosophical analysis of science than to its practice, for whom publication had in recent years meant encyclopedia articles, popular addresses, and memoirs of departed colleagues. With the death of Powell and, excepting the Harriman Expedition report, with no major paper for nearly a decade, Gilbert seemed almost to belong to a former era. Scientific honors rained upon him, but they paid homage to the author of the *Henry Mountains* and *Lake Bonneville*, not to the scholar of the High Plains, glaciation, or lunar geomorphology. During this period he held executive offices in the Cosmos Club (1904); in the Washington Academy of Sciences (1898), an umbrella institution to oversee the proliferating scientific societies in the Washington area; in the American Association for the Advancement of Science (1900), being president when it adopted *Science* as its official periodical; in the Geological Society of Washington (1893); in the National Geographic Society (1891–1900), as a member of its board of managers; in the Philosophical Society of Washington (1893); and in the Geological Society of America (1893)—and, as a result of reelection in 1909, he became the only man in its history to hold the presidency of the society twice. He trained several young Survey geologists in field methods and corresponded with many others, especially commending them on recent achievements—perhaps these were his substitute for the students collecting around Chamberlin, Davis, and Le Conte. His long affiliation with the Survey set high standards for it and made Gilbert one of the formative figures of the federal scientific establishment. By 1902, however, it seemed that he was ebbing from the active social and intellectual shores of geology.

But there was far more to Gilbert than that. Like the great masses beneath Lake Bonneville, so long depressed by an overburden of water, there remained an elasticity. When the climate changed, when the burdens and frustrations of the past years finally evaporated, he began a dramatic rebound. His final years, even though broken by illness, were a ringing testimony to his continued powers, to his blend of patience and intelligence. While he became more dignified as the years progressed, he never became a simple dignitary. The Survey which he had done so much to shape and direct repaid its debt by sending him back to the Far West to revive the themes of his past. If *Lake Bonneville* had been a culmination of American geology of the nineteenth century, Gilbert's hydraulic studies in California were an inauguration to the twentieth.

6. The Inculcation of Grove Karl Gilbert

He so wisely conserved his failing health and so regulated his working hours that the quality of his product was not impaired. . . . His last paper is an amplification and clarification of one which he wrote more than half a century ago. The cycle is now complete.

—*George Otis Smith*

Geophysics in the Giant Forest

By 1903, Gilbert had begun reconsolidating his personal life and his scientific career. He elaborated a new network of social ties to compensate for those he had lost and redirected his scientific curiosity to a new landscape. But, just as his new social environment recapitulated old arrangements, so his new terrain held old themes. Despite a near fatal illness in 1909, he channeled the intellectual vigor of these last fifteen years of life into a grand summary of the inquiries which had dominated his entire career, putting into final form his observations and interpretations in glacial and structural geology, hydraulics, and geomorphology.

In 1900, at the age of fifty-seven, Gilbert had been awarded the Wollaston Medal of the Geological Society of London—Britain's most coveted prize in geological science. He was the third American so honored; only James Hall and James Dwight Dana had preceded him. Yet, with that touch of whimsy which ever balanced his conservative temperament, he wrote a friend that the thing he "was proudest of" was the prize he had "won at progressive euchre" that week. Perhaps the facetiousness was apropos: the Wollaston Medal honored only the contributions he had made over the previous thirty years. It looked to his past while Gilbert, even amid the doldrums he experienced at the turn of the century, preferred to live in the present. By the end of 1903, he was prepared to look to the future.[1]

By 1903, Gilbert had discovered surrogates for his diminishing family ties, for the friendships that had revolved around Powell, and for active membership in various scientific societies. He took lodging with the family of C. Hart Merriam—whenever he was in Washington, which amounted to about six months a year, he boarded in the two upstairs rooms that the Merriams had specially prepared for him. One of the country's foremost naturalists, chief of the Biological Survey, and prominent self-taught anthropologist, Merriam easily assumed the intellectual and social role Powell had formerly occupied. Active in Washington's social and political arenas, Merriam gave Gilbert access to a wider audience than he would have known if left to his own means, while at the same time furnishing him with a domestic environment akin to the Nutshell. Although he steadily withdrew from active participation in many scientific societies, holding mostly honorary offices, Gilbert frequently accompanied Merriam to meetings like those at Gifford Pinchot's house, where Theodore Roosevelt spoke on forest conservation, and to Alexander Graham Bell's weekly gatherings for the National Geographic Society's board of managers (both were members).[2]

In return, GK participated rather fully in the domestic habits of the Merriams. He often reciprocated with entertainments of his own. Occasionally he brought in special performers, such as magicians; commonly he read aloud to the Merriam children at night. At times he accompanied the family to the theater or the circus. And he could often be counted on to share the burden of tiresome houseguests, like John Muir. "Muir has gone," he sighed to Arch, "and I think everyone is relieved. He was a hard guest to entertain because he was very dependent. He did not take the trouble to master the street systems and expected a companion wherever he went. And his talk, which never ceased, became very tedious." He added: "I don't believe I have any use for him."[3]

In like manner, Gilbert's friendship with J. H. Comstock, a professor of geology at Cornell, ripened after brother Roy died in 1901. By simply transferring his base of operations from Rochester to Ithaca, Gilbert continued his forays into the landscape of upstate New York. So, too, did he shift his headquarters for research in far western geology. In 1903 he began his personal exploration of the Sierra Nevada; in 1905, the study of its devastation by hydraulic mining became an official assignment. Gilbert set up quarters at the University of California at Berkeley in "a very comfortable room in the Club." He found the existence pleasant, though he wrote to a friend that "the typical Club-man would find it dull." In effect the faculty club—with its radius of connections throughout the Bay Area—substituted for his subsiding participation in formal scientific societies. Where Gilbert had formerly played whist and billiards at the Cosmos Club, he now played at Berkeley; where he had earlier lunched with the Great Basin Mess, he now ate overlooking the shores of San Francisco Bay.[4]

Gilbert's first excursion into the Sierras came in 1903. The journey was officially endorsed as a pleasure trip by the Sierra Club which he had joined, but it provided his first semiscientific introduction to Sierra geology. He had briefly examined the eastern front of the Sierras near Mono Lake when, as chief of the Division of the Great Basin, he toured sites of Pleistocene lake beds there with I. C. Russell in the 1880s. Otherwise he knew the landscape only through its literature. Clarence King's and John Muir's lush descriptions of Sierra landscape had long been favorite reading for Gilbert, and on personal inspection he found that their enthusiastic prose, if anything, verged on understatement. Once again "astride the occidental mule," stimulated by a glacial topography that was novel compared to that he had known in Alaskan fjords and along Great Lakes shorelines, Gilbert experienced a scientific and social reincarnation.

The party traversed the High Sierras around Mount Whitney and the Kern River—King's old haunts. GK found the country and the company equally satisfying. A second expedition was organized on the heels of the first—this one to Yosemite, Muir's preserve. Merriam, a member of both parties, recorded a typical August day. They were "up at 5, ate breakfast, and at 6 set out on horseback for the summit of Conness (12,500 ft.)." They arrived a few hours later, riding most of the way to an old Coast Survey camp. "The air was clear, the smoke-haze not having risen yet, so we had a glorious outlook in all directions." They left the peak in midmorning, "lunched in the basin where we had camp last night (Partition Pond) at noon, stopped and photographed along the way, and reached camp on Budd Creek in Tuolumne Meadows before 4 PM." Nights were cold but, as Merriam scribbled into his journal, "we have a big fire of dead pines—and our beds under live ones."[5]

Also in 1903, en route between his western work and his Washington office, Gilbert conducted his habitual spur trips to points of interest. Besides layovers in Jackson, Michigan, to visit his sister and in Chicago for a chat with Chamberlin, he monitored the water gauges installed by the Lake Survey on the shores of Lakes Michigan and Huron—further checks on his theory of isostasy—and reviewed

Gilbert at Yosemite, 1903. *Courtesy of the USGS Photographic Library.*

observations in Colorado. From Washington, still hoping to convince the Carnegie Institution to implement his scheme to bore into the crust, he visited the prospective site near Lithonia, Georgia. From Washington, too, he examined the nearby Glen Echo quarry for its pattern of jointing and stresses, a laboratory for what he had observed in the Sierra domes. In the same way he quickly surveyed Stone Mountain in Georgia, a granite dome. Before the year ended, he was chairing a committee to revise the rules for stratigraphic nomenclature adopted for Survey atlases—rules he had designed in 1889 for Powell. As the Survey's annual report remarked, "The work of the committee this year involved the establishment of important precedents." It was work Gilbert engaged in for basically conservative reasons, namely, to stifle confining classification systems. But Walcott was right: the two committees Gilbert chaired set precedents whose influence continues to the present day.[6]

The Sierra Club repeated its journey to the Sierras the following year. This time the party included some illustrious California geologists, especially A. C. Lawson who had succeeded Le Conte as head of the geology department at Berkeley. Speculating on the relative importance of glacial and fluvial erosions, the meaning of the special terraces staggering the long eastern slope of the Sierras, the origin of the granite domes peculiar to the granite crest of the Sierras, and especially (for Gilbert) the distribution of stresses as manifest in jointing patterns, Lawson and Gilbert exchanged ideas freely. When they decided to publish, Gilbert deferred to Lawson, whose *Pleistocene Topography of the Sierra Nevada* appeared later in 1904.[7] GK's own observations, bolstered by further summer excursions in 1907 and 1908 into the High Sierras, coagulated around three broad themes: the mechanisms of glacial erosion, the relationship of exfoliation—jointing structure parallel to the surface—to topography, and the cause of rounded hilltops. He produced terse essays—some were even published in Sierra Club bulletins—brought out over a period of ten years. But in spite of their brevity several became immediate classics.

Gilbert supplemented his trips to the Sierras with excursions with the International Geographic Congress to Niagara and to the Louisiana Purchase Exposition in St. Louis. Again he visited Chicago for conferences with Chamberlin and Charles Van Hise, the petrologist and author of one of the major conservation tracts to emerge from the Progressive Era, *The Conservation of the Natural Resources of the United States.* En route to the Sierras, he briefly scanned El Solitario, a dome near the Big Bend region in Texas. The locale was an opportune one, for it combined two prominent themes of Gilbert's structural geology: the laccolith and the Basin Range structure. Placed on the eastern fringe of the Basin Range province, it nicely complemented Gilbert's inspection of similar forms in Oregon and southern California. When he left the Sierras several weeks later, he turned north for another look at the Cascades of the Columbia, another familiar junket to his gauges on the Great Lakes, and the inauguration of a final study of Niagara.

But these pleasuring excursions, brief as they were, also amounted to a preliminary reconnaissance of the California landscape. They raised Gilbert's imagination once more to its kindling temperature. Ignition came in 1905 with a topic whose investigation not only focused his impressionistic knowledge of Sierra geology but integrated the dominant interests of a lifetime. That subject was hydraulic mining.

Hydraulic mining began with the first panwashed gold from the gravel placers in Sierra streams. Because of their great density, gold nuggets collected in eddies and pockets of reduced velocity in the streams, while finer material washed away. This same principle guided hydraulic mining in its industrial form. The technology of processing the ore evolved rapidly after 1848. The rocker replaced the pan, the sluice superseded the rocker, and ultimately the stream itself was replaced by vast networks of flumes, hoses, and nozzles that detonated hillslopes with concentrated blasts of water. The decisive inventions in the advancement of hydraulic technology came in 1852 and 1853, but the new industry stalled within a few years because of uncertain water supplies and a national economic depression. By 1864, however, spurred by the lucrative bonanzas unearthed at the Comstock, San Francisco entrepreneurs accelerated their efforts to flush easy wealth out of the western slopes of the Sierras. After nearly a decade of economic slump, California mining boomed again as full-scale hydraulicking, a corporate enterprise with heavy capitalization. Hydraulicking was more efficient than small placer claims worked manually by partnerships and more profitable than hard-rock mining. Interestingly enough, the fortunes of hydraulic companies varied as much with seasonal rainfall as with the market value of gold; those who owned water and reservoirs eventually bought out most of the mines. The water monopolies Powell had warned about with respect to agriculture were equally applicable to mining. The greatest Sierra resource was its water—a point that, until Gilbert made it unblinkingly clear in 1917, was commonly neglected.[8]

Hydraulic mining worked because large volumes of Sierra gold were stored in ancient placers, vast beds of gravel originally deposited by Tertiary river channels and subsequently elevated by tectonic activity into the sloping profile of the western Sierras. These buried reservoirs were the source for the natural placers of streams like the Yuba which drained them. Hydraulic mining expanded and accelerated the natural process by flushing away enormous slopes of the auriferous gravels, after passing them through sluices which, with incredible waste, trapped some of the gold. When certain tiny valleys could no longer absorb the scale of slope wash, tunnels were drilled to export the tailings into a larger river system. Hydraulicking was strip mining at its most rapacious.

By the late 1870s, large corporations dominated the business. Huge fires illuminated the mine sites and allowed for round-the-clock operations; telegraph lines coordinated activities; the nozzles became literal cannons, like the Little Giant, with a six- to ten-inch bore; and superior drills made tunneling more efficient. In 1873 an astonished Rossiter Raymond, a government mining engineer and crony of King's, reported in full bureaucratic understatement that "in fact hydraulic mining has assumed proportions heretofore scarcely dreamed of." In 1879 the North Bloomfield Mine awed reporters by the "real pleasure, very distinct, but hard to describe, about this gigantic force." Small wonder that Clarence King—himself an accomplished mining engineer—blasted American mining that same year for its extravagance and harangued Congress to give the new U.S. Geological Survey some powers of indirect regulation. By then the industry claimed over a thousand miles of ditches siphoning streams into the muzzles of hydraulic armaments—and a steady tide of debris slowly tumbled into the valleys below. By 1880 some 47,000 acres of rich farmland in the Sacramento Valley had been damaged or destroyed by the rising flood of tailings. Cities like Marysville on the Yuba—the most abused river of the lot—built dikes along riverbanks. The bed of the Yuba, however, merely continued rising at a rate of almost a foot a year and, as the levees and dikes strove to keep pace with it, the river became a virtual aqueduct. In 1875 and 1876, the levees surrounding Marysville ruptured and the town was inundated.[9]

Agricultural interests rallied behind the Anti-debris Association, incorporated in 1878. But it was not until 1884 that they successfully won a legal suit, an injunction against further mining. In 1893 Congress established the California Debris Commission to regulate the industry, which it almost totally shut down by stringent licensing. The key to the legal and political war over hydraulicking lay with the commercial interests of San Francisco, who, by the 1880s, knew that the economic future of the state resided in agriculture rather than in mining. The decision makers had, as well, two engineering reports to work from: an 1879 report by Lt. Col. George Mendell of the Army Corps of Engineers and an 1880 report by William Hall, the state engineer.

Meanwhile, as a result of the Caminetti Act of 1893, which set up the Debris Commission, an ambitious system of engineering works came into existence with the intention of helping agriculture. The scheme included shoring up river levees, constructing debris dams, reclaiming swamplands, and helping accelerate the discharge of debris through the Sacramento channels. The Sacramento thus became the third American river to be managed by a commission of engineers, the others being the Mississippi and the Missouri.

In 1904, responding to further engineering proposals, with the price of gold high and the value of wheat low, the mining interests organized to repeal the legislation against them. Their campaign included a memorial to President Theodore Roosevelt to have the U.S. Geological Survey reexamine the evidence. Roosevelt agreed, and in 1905 the Survey handed the assignment to its most illustrious investigator of streams: G. K. Gilbert.

Mining evidence was still in abundance—so much so that by 1904 stabilization projects, mostly dams and levees, were just cranking into high gear along the Yuba. In the short span of thirty-five years, some 1,295,000,000 cubic yards of debris had been washed from Sierra slopes. For Gilbert the debacle was a superlative chance to synthesize investigations which reached back to the Powell Survey. In the natural and engineered transportation of this debris load, he could study bedforms, channels, shorelines and terraces, the fluvial transportation of sediment, and the modification of landscape induced by human settlement. Since it dealt with streams, the study partook of that unifying thread, water, which ever since Powell's *Arid Lands* report had controlled conservation thinking. Finally, the study was both pure and practical science: pure in that it blended laboratory and field in a search for a theoretical explanation of fluvial transport, practical in that it had direct social-political consequences. It is no accident that GK launched his study of this debris flood at the same time that he was concluding his final treatise on Niagara, for he conceived of both the Sacramento and the Niagara as physiographic engines performing work. The Sacramento, a product of human engineering, demonstrated the power of aggradation; Niagara, its natural counterpart, epitomized fluvial degradation.

Gilbert quickly arranged his investigation. Little more than a rapid reconnaissance was initially required, for the uprooted Sierra landscape was still fresh in his memory. Yet his research pointed in a direction he had been longing for ever since he attempted to sketch a logical model of stream behavior in the *Henry Mountains*: the study, in short, demanded a quantitative, precisely instrumented analysis of sediment transport. He established field headquarters in Sacramento, a place "so dull," he wrote in mock despair, that he had "downslidden to billiards in a public billiard room." From this base camp he forayed to nearby stream channels, reservoirs, dams and levees, tailing shoals, and mines. He found most of the maps unsatisfactory and most of the earlier estimates on the quantity of the debris disappointingly erroneous. Over the course of four years, from 1905 through 1908, he meticulously revised the previous estimates of the Army Corps of Engineers and the California Debris Commission.[10]

Once the debris was quantitatively measured, the next problem involved an investigation of how all this debris would be moved, both under natural conditions and in accordance with existing or proposed engineering works. Gilbert addressed this second phase of the debris problem by constructing a set of flumes to painstakingly measure the relative influences of different variables in stream transport. He erected his flumes, the first hydraulic laboratory of the Geological Survey, on the Berkeley campus in 1908. After designing the experimental model and testing the apparatus, he left most of the experimental runs to an assistant, E. C. Murphy. The completed results were successfully published in two major monographs, to be discussed below: the laboratory work in 1914, in *The Transportation of Debris by Running Water*, and the fieldwork in 1917, in *Hydraulic-Mining Debris in the Sierra Nevada*.

The entire investigation took over ten years. There was little urgency, since only bootleg mining had continued in the mountains over the previous twenty years, and GK's inquiry was less an anatomy of a landscape in danger than an autopsy of one deceased. More directly responsible for the long preparatory and experimental period, however, were two quite serious interruptions—one for a public and one for a personal catastrophe. In 1906 San Francisco was nearly decimated by an earthquake, and in 1909 Gilbert suffered a near fatal illness which lingered with disabling effects for years.

The former somewhat perversely delighted him. "It is the natural and legitimate ambition of a properly constituted geologist," he exulted, "to see a glacier, witness an eruption, and feel an earthquake. . . . When, therefore, I was awakened in Berkeley on the eighteenth of April last by a tumult of motions and noises, it was with unalloyed pleasure that I became aware that a vigorous earthquake was in progress." Calmly he timed the intervals between tremors and measured the direction of ground waves. For some time, faint aftertremors persisted. A column of smoke rose over the city. Dynamite could be heard booming across the bay. For forty-eight hours San Francisco was isolated from outside traffic, then the ferry reopened with Gilbert on it. He fired off a telegram to Merriam, telling of friends spared and "burned out." "The westward progress of the fire north of Market Street," he jotted into his notebook,

> has been checked chiefly by backfiring at Van Ness Avenue.
> The houses opposite were blistered and had glass broken, and
> at one place the fire broke across, to be checked at Franklin St.
> Backfiring is in progress N. of Pac. Avenue, and apparently
> being carried to the waterfront. From this the fire rushes up
> the slope of Russian Hill, consuming block after block of
> houses—chiefly of wood. The flames work with wonderful
> speed. While I lingered, whole squares were consumed. An
> hour is probably enough to raze a square of wooden houses.

Unruffled by the spectacle, he calmly measured the burning time for a two-story house: "roof gone in 7'; first falling of wall in 9'; flaming ruins in 12'."[11]

Shortly afterward, Gilbert was appointed to a state commission, charged with investigating the catastrophe, and he was also assigned to a later research team organized by the U.S. Geological Survey. Both groups published results in 1907. Gilbert's formal contribution was similar to that which he gave for glaciers in the Harriman report: it was largely descriptive and pictorial, although in certain popular articles he succinctly analyzed the mechanics of stress which led to earthquakes.

The teams of geologists and engineers who investigated the earthquake divided the field both geographically and thematically. Gilbert worked mostly north of the Golden Gate. At the same time, he continued his pattern of summer vacations. In past decades, these would have taken him to Niagara; his new environment directed him toward the Sierra Nevada. He organized informal touring expeditions in 1907 and 1908. The first was a private "extravagance," as he boyishly called it, in which he conducted a small party of close friends on a pleasuring trip. The party answered a charming invitation. "You are cordially invited to my house party in August," GK wrote with mock formality. "My cellar is the Yosemite Valley, my drawing room the Tuolumne Meadows, my attic Mono pass, and my staircase the Tioga road." His sister, Emma Loomis, answered the announcement, along with the Comstocks from Cornell and Alice Eastwood, a botanist friend from the California Academy of Sciences. Gilbert outfitted the group with two packers, a cook, and sixteen horses. At 6 feet 1½ inches, still robust from years of field campaigning, his reddish hair graying but closely trimmed, Gilbert was a quiet but commanding presence. Among the party he soon became known as Charlemagne. It was a leisurely, somnolent month, with warm days and cool nights, the laughter of old friends, and miles punctuated by frequent stops for instruction and appreciation. Gilbert skillfully cashed in his rich experience as an exploration geologist for the pleasure of his friends.[12]

Gilbert's second summer in the Sierras was rather more scientific: he escorted his old comrade Willard Johnson and a visiting geologist from Australia, E. C. Andrews. Some records remain for the trip. A mutual fascination with glacial topography had brought Gilbert and Andrews into contact. Begun as a study in comparative glaciation, the expedition merged imperceptibly into a grand reconnaissance of the landscape. A few years later, Andrews recalled the experience in a letter to William Colby, president of the Sierra Club. "In the midst of magnificent scenery and in company with one of your noblest natures, I learned to love every inch of the great Sierras," he wrote.

The man I refer to is Dr. G. K. Gilbert. . . . He it was who taught me the names of all your forest trees and their geographical distribution. He it was who read to us of the lives of Galen Clark, John Muir in the Sierras and of John Wesley Powell in the Grand Canyon.[13]

Gilbert's pastimes, as Andrews later observed, were mainly cerebral. Except for occasions when he read aloud, he continually indulged in intellectual exercises. If a horse or mule fell out on a trip, a pack broke out of its diamond hitch, or the party halted for a meal,

Gilbert would ask at once for a "problem" to be given him to solve. The writer's stock of questions of maxima and minima, on astronomy, on motion round curves, on inertia, on flywheels, on nodes of curves, on physics, were soon used up, as Gilbert could see through a problem very quickly. In return he would always propound a difficult problem.

At night, Andrews remembered, "in the sleeping bags, he would teach us the names of all the principle stars, constellations, and so on."[14]

Another favorite recreation inherited from his youth was a fondness for impromptu or round-robin poetry composition. Usually Gilbert dictated the rhythm and the topics, but in 1908 he found himself the victim. To a flowery line from Gilbert, Andrews replied:

Grove Gilbert was our captain bold
Of Henry Mountain fame
He lit his torch with lac-o-lite,
And straightway made his name!
His many "faults" were mighty ones,
No common, garden brand.
His Wasatch "slip" in Mormon land
Is known on every hand.[15]

In Washington, Gilbert's habits of leisure and society likewise continued early tendencies. He loved walking, and, despite his age, tramped whenever possible. He rowed on the Potomac nearly every afternoon. "At other times," recalled Merriam,

if like-minded companions were available, he played billiards, dominoes, or cards, or read aloud; and when alone, alternated reading and solitaire. Once or twice a year he went to see a

game of ball, or took the children of some friend to the circus;
but he did not care much for the theatre or for music, and
needed the stimulus of companionship to indulge in either.

He withdrew increasingly from public meetings, including those of
scientific societies. Even the weekly trips to Bell's house could
make him uneasy. "The interesting things" Bell told him, Gilbert
wrote to Arch, "make me feel my social limitations." Reading aloud
could at times be no less uncomfortable. Finishing a sentimental
novel, he more than once reached for a handkerchief, while "the
women only laughed—at us," he lamented. The episode was hardly
an isolated case; the basic dynamics of GK's personality were little
different from those of the young husband who asked Henshaw to
escort his wife to the plays she wished to see.[16]

To those who knew him well within the confines of his surrogates for the Nutshell, he was warm, unpretentious, direct, his
humor and speech somewhat unconventional. He had little use for
formal society. Yet to those outside his sheltered environments he
must have seemed quite different, more like the man recorded in
those careful photographic portraits—a little stiff yet self-conscious,
vaguely preposterous with his meticulously self-repaired clothes,
occasionally brusque manner, and dogged sense of responsibility.
Outsiders saw the formidable author of the *Henry Mountains* and
Lake Bonneville as composed and fixed as a set of axioms. Such was
his talent for self-effacement that few saw that vulnerability which
made such controls imperative and that acute inner sense of "social
limitations." Yet to both groups it was apparent that, whatever his
situation, Gilbert surrendered neither his dignity nor his modesty.

In March 1909 the tempo of his life was abruptly shattered by
illness. The symptoms developed in Berkeley. They were diagnosed
as apoplexy, and "while my physician's tone is optimistic," Gilbert
wrote the Washington office, "my own impression is that my general physical condition is permanently lowered and that any later
change will not be in the way of improvement." To Arch he confided, "You will be surprised to learn that I am quitting California,
and don't expect to return." He contracted his scope of interests to
the most prominent and advanced studies—the debris transport
problem. His only complaint was that he found it "a grind to have to
think so much about myself as I have recently."[17]

After a couple of weeks, he evacuated Berkeley for his sister's
house in Jackson, Michigan. Since Murphy did most of the laboratory work, there was little loss on that score; he had "planned more
field work" on the debris problem, "but it is not essential," he ex-

plained to Arch. The investigation had advanced "to such a stage that it can go to report without material loss." Nevertheless, the Survey dispatched W. C. Mendenhall to take notes on the project so that its work might not vaporize if Gilbert died. For a week GK dictated from memory, but so exhausting was the procedure that he could speak only for fifteen or twenty minutes before requiring an hour or so of rest. The hundred pages of notes testify as much to his patience, perseverance, and self-discipline as to the condition of mining debris in the Sierras.[18]

With pitiless slowness his health improved, and to aid it he followed a rigid schedule of leisure, a demanding chore. When the summer was over, he took up residence again with the Merriams in Washington. "My niche and my rut were both ready for me and I'm settled in them . . . plodding and loafing in the same old way." The following summer of 1910, he stayed with the Comstocks at their summer cottage on Cayuga Lake. He found enough strength to outline chapters for the debris study, but the work went slowly: not only was his strength at low ebb but, as he put it, "I've had to change my point of view repeatedly and go back and rewrite." Back in Washington again, he completed his slow withdrawal from formal organizations and society functions. The National Academy of Sciences held its meetings, but Gilbert "hardly saw a member." Even informal dinners brought distress: "There were ten at the supper and the conversation used me up, so that I fled before the party was half over." He delivered his medals ("$300 worth of gold") to the National Museum. In addition to the Wollaston, there was the gold medal of the National Geographic Society, awarded in 1909, and the Daly Medal from the American Geographical Society of New York, awarded in 1910. When the Walker Grand Prize in Natural History was awarded to him, he immediately turned over the $1,000 check ("it literally took my breath away," he told a friend) to a young geologist for continuing studies on the San Andreas fault zone. Alarmed over his precarious health, GK feared lest "the money be lost to science." He had already disposed of most of his geology books: when the library at Denison University in Ohio tragically burned, he had donated much of his own collection to restock it. The geology department chairman, Frank Carney, assured the old man that "the Gilbert Library is a busy place about every hour of the day." On the fiftieth anniversary of his graduation, Gilbert reluctantly declined an offer to deliver the commencement address at the University of Rochester. He was concentrating all his limited energies on the debris problem.[19]

By the summer of 1911, a stoic regimen of exercise and disciplined leisure had slowly restored his health. With this restoration of equilibrium, he could again "pursue 'the even tenor' of my way." That summer and the following one, he stayed at a small hotel in Annisquam, Massachusetts, on the New England coast. "For the first time in many years," he exulted to Arch, "I am living in view of the sunset." He exercised in a dory early in the morning. "I seem to be the only one besides the fishermen and it is a bit lonesome," he noted dismally. He even tried "taking a widow along . . . but she talked too much, and I don't think I'll give her another chance." Later he discovered an admirable companion in an old acquaintance, Charles Van Hise, then president of the University of Wisconsin, and together they "rode on a full grown Atlantic swell which was making magnificent surf at the shore, a most exhilarating experience." His tireless, penetrating mind demanded mental exercise on an equal scale—he found it in a "study of sand ripples . . . a good theme for boat excursions and . . . somewhat related to my laboratory data." And it was that hydraulic data transported from California which he began to rework, like the ceaseless surf at Annisquam, until he expanded his work schedule to nearly a full eight hours a day. This required extraordinarily close attention, aggravated when, as he explained to Arch, Murphy "turned in some drawings that showed he lacked the fundamental conception of the contour— and I went back to the original data."[20]

Otherwise Gilbert spent his final years according to a familiar itinerary—winters with the Merriams; summers with his sister in Michigan; precious research months at Berkeley, San Francisco, the Sierras, and ultimately the Great Basin. His habits were still largely domestic and clubbish; he laughed at his meticulous passion for billiards. With his renewed health, he gradually expanded his range of geologic endeavors. Even when virtually bedridden, he had undertaken a long debate with Bailey Willis on the nature of the earth's interior and on continental problems in structural geology. Willis would leave a written question beside Gilbert's bed every day or so, and Gilbert would write out a reply; despite the awkward format, the two embroidered their notes into a lengthy symposium. Soon the circle included Joseph Barrell, who greatly admired the elder statesman of geology. The correspondence found its way into GK's final assessment of isostasy, published in 1914. There were more informal papers on topics like the formation of ripple marks, and, of course, looming in the background like the Sierra crest line, there were the hydraulic studies.[21]

On these he worked indefatigably. The volume of computation required for the reports was staggering; even Gilbert was "rather dismayed" by it. In an age without calculating machines, the work was tedious and exacting. At times he had the loan of an assistant. Mostly he worked by himself, however, shunning even the slide rule in favor of the more familiar logarithmic tables. "But this present work," he noted with relief, "has introduced me to log. section paper, and that has become quite familiar. Of course it involves the same principles as the slide rule." Several times he stalled in the morass of equations; once, stymied on a problem involving the "mathematics of mean values," he turned successfully to an article in *The Century Cyclopedia* written by C. S. Peirce. At other times, as he had so often in the past, he corresponded with Robert Woodward.[22]

The work demanded, too, an abnormal volume of reading, especially of texts by European authors. Here François Matthes helped. This amount of reading was an unusual exercise for Gilbert, who had habitually worked on geologic questions more or less in isolation. The translated works involved physics and engineering (that is, hydraulics and hydrodynamics) more than geology, and that may be the reason. Most of Gilbert's prior knowledge in these fields probably came from sections of general American engineering texts, passages in Rankine, and, of course, Andrew Humphreys and Henry Abbot's classic, *Physics and Hydraulics of the Mississippi River.* The latter, published in 1861—the same year as Newberry's report on the Colorado River for the Ives Expedition—summarized much of the European literature of its day. In 1913, well after his flume experiments were completed, Gilbert remarked to Arch that he was "still busy with the libraries, and found the French engineers have done some good work I did not know of. Glad I did not go to print without finding out as to it." Yet he no more stood in awe of those monuments of French science than he did before the Arc de Triomphe. The Yankee in him rebelled. "Shall know right smart of hydraulics and some of hydrodynamics when I get thru," he concluded. The monograph, a summary of his laboratory work, was completed in June 1913, on Friday the thirteenth. "A lucky day," Gilbert felt. Its title: *The Transportation of Debris by Running Water.* It was printed the following year.[23]

Immediately after submitting it, GK went back to the field to reexamine hydraulic debris in nature. His contact with the Sierras once again rekindled his imagination, and he satisfactorily resolved two nagging questions: the relation of exfoliation to topography and the cause for the convexity of hilltops. Both analyses are among his most celebrated short essays. The studies emerged somewhat serendipitously in the course of his tours of numerous mines. Along the

Yuba River, he examined closely the consequences of a major flood in 1906. Numerous calculations were demanded in the compilation of his field data and, just as he had turned to European scientists for assistance in hydraulics, so he requisitioned aid from such sister agencies as the Coast and Geodetic Survey and the Navy. In particular, he needed information on the tidal movements in San Francisco Bay. The "tide problems," he wrote excitedly to Arch, "continue to be fascinating."[24]

He transferred quarters from Berkeley to the University Club in San Francisco; now when he visited the Berkeley faculty club, it was to borrow books and play billiards and bridge. By March 1915, his latest monograph was in galley proofs. Gilbert undertook the painstaking enterprise of proofreading. "Most of the things I find to correct are slips of typists or compositors, but some of them are my own," he lamented, "and that's saddening for I went over the manuscript many times. I cannot hope to eliminate all now, but am doing my best."[25] *Hydraulic-Mining Debris in the Sierra Nevada* was finally published two years later. Gilbert was seventy-four years old, but even at the time of printing he was in Utah, indulging a lifetime's concern with Basin Range structure.

When he assaulted the mountains of the Great Basin for the last time, his health was measured but precarious—he wrote that it was "holding its own and a little more." At times of overexertion, he could collapse; in controlled moderation, outfitted with a motorcar, he found fieldwork bearable, even delicious. In 1914 he had spent several weeks bivouacked along the Wasatch fault. In 1916, with his hydraulic studies at last terminated, he launched another determined campaign against the Basin Range. Because his maps had been lost in 1902, he had been unable to systematically answer his critics, particularly J. E. Spurr and C. R. Keyes, while at the same time much of what he intended to say had appeared as independent discoveries by others, notably George Louderback and William Morris Davis. But it was a study he was especially competent, and anxious, to prosecute. In Arch's car, accompanied by Willard Johnson, he began surveying scarps—in the Klamath Falls region of Oregon, in Kern Canyon of the Sierras, and in northern Nevada near Winnemucca. As he expressed it, "The mountains, and especially their bases, have stories to tell me."[26]

In 1917 he revisited the Wasatch for detailed measurements. He followed with another excursion to the House Range. But this time he lacked his old comrade, Johnson. Tormented by deteriorating physical health, and with his major work—*The High Plains*—well behind him, Johnson committed suicide in the Survey's Washington office. Gilbert solicited the aid of a Forest Service official along with François Matthes—both friends of Johnson's—"in the prospect of rescuing a portion of Mr. Johnson's results from the general wreck of his unfortunate life." Johnson had been the last nexus with his youthful explorations; Dutton had died in 1912, Howell a year before that.[27]

Studies of Basin-Range Structure, the product of this, his last expedition, was published posthumously in 1928, fifty-seven years after he began his first western survey by stepping off a train into a panorama of desert peaks and playas at Halleck Station, Nevada.

Gilbert the Glaciologist

Despite his frequently debilitating health, GK's last years were charged with an unrivaled intellectual vigor. Forced to shrink his range of associations, both professionally and personally, and to constrain his patient curiosity, he nevertheless managed to summarize his scientific career with articles on all his major geologic concerns—geomorphology, glaciation, structural geology, and hydraulics. At the same time, he continued to write memoirs of scientific friends who had died, and, as if to anticipate his own, he arranged to have his entire set of publications bound for storage. This resulted in thirteen volumes.

Francis Vaughn, a student of A. C. Lawson's at Berkeley, described Gilbert's status during these final years. "Not all the assignments by Lawson," Vaughn recalled, "were meant to be torn apart under critical study. A few works were intended as models to show us how beautifully some of the problems and presentations in geology could be treated. The two most important of these were *Lake Bonneville* and the *Geology of the Henry Mountains* by G. K. Gilbert." Needless to add, "Lawson had a profound admiration for Gilbert and succeeded rather well in passing this reverence along to us." To the Berkeley students, Gilbert "was a singularly mild-mannered man and about as quiet as a shadow"; they were "somewhat amazed by the way the dynamic Lawson dropped everything and deferred to him whenever he put in his appearance at Bacon Hall." But they "did not delay in doing the same ourselves. A perplexed brow, when the old gentleman was scanning a shelf of books or papers, was sufficient to cause someone of us to jump and offer assistance." Actually that shadow of a man needed little help. He was determinedly but privately putting his own works in order.[28]

One important set of studies pertained to the general geology of the Sierras, including brief papers on the mechanism of glacial erosion, exfoliation, and the convexity of hilltops. The Sierras confronted Gilbert with a glaciated landscape different from those he had investigated before. In New York and Ohio, he had encountered the evidence of continental glaciers, vast ice sheets that had ground southward across broad but relatively flat expanses of country. He was familiar with these terminal moraines and drumlins, outwash plains and effects on lake levels, ice dams and outlet channels. The situation in California was closer to, although still different from, that Gilbert met on the Harriman expedition—the salient geomorphic puzzle of the tidal glaciers in Alaska was to discriminate between fluvial and glacial topography along the coast. In the Sierras he encountered valley glaciers and cirques, but the major problem was identical—to discriminate between the erosional patterns left in the mountains by fluvial processes and those left by glacial processes. In this sort of observation, however, he excelled. The resolution of the landscapes into two fundamentally distinctive regimes was as essential to his intellectual perspicacity as to the construction of his sentences. Similarly, the riddle of erosion beneath a tidal glacier led to considerations on the mechanics of abrasion and plucking by valley glaciers. The stress caused by glacial weight, as with the waters of Lake Bonneville, brought Gilbert to the question of crustal elasticity, with respect both to the distribution and cause of joint patterns and, in turn, to the problem of exfoliation. Finally, the massive granitic domes so characteristic of Sierra sculpture, as well as the tiny mounds of gravel tailings from hydraulic mines, forced him to consider various mechanisms for generating convex hilltops.

GK's "Variations in Sierra Glaciers," published in 1904–05, and "Moulin Work Under Glaciers," published in 1906, are descriptive reports identifying unusual forms. The work on the "glacial mill," or moulin—the stream of meltwater flowing from the glacial terminus—is interesting as a point of observation, for in it he returned to the subject which had launched his career: potholes. "Variations in Sierra Glaciers," in the same way, returned him to the themes of his first professional essay by making an analogy between glacial termini and fluctuating lake levels. He concluded by asking for the assistance of Sierra Club members in documenting the position of the glacial termini as a measure of climatic oscillations.

A more sophisticated analysis characterized the 1906 "Crescentic Gouges on Glaciated Surfaces." Here Gilbert furnished a mechanical cause in the form of a stress analysis for a particular form of "rock fracture" associated "with striae and other evidences of glacial abrasion." The crescentic gouge had a shape similar to that of the chatter mark identified by Chamberlin, but it had a different

The scientific Sierras: exfoliation domes, large and small, of the Tuolumne Basin at Yosemite. Gilbert's 1903 photograph should be considered—especially from this perspective—with his almost obsessive photography of other curved landforms, such as ripple marks, dunes, and gravel tailings. *Courtesy of the USGS Photographic Library.*

genesis—GK imagined the gouge as geometrically and physically related to the "conoid of percussion." While the conoid is normally the product of a sharp blow, he conceived an alternative force in the differential pressure of a glacier acting on a boulder, where the boulder is positioned on a prominence and cushioned by sand and other basal debris. Actually, rupture occurred along two stress planes—one was the conoid of percussion caused by compressive stress exerted downward or obliquely, the other a tensile stress which produced a vertical face around the crescent. The cause for this latter phenomenon, he suggested, was "differential friction":

> It varies with the material of the two bodies in contact and is directly proportional to the force, normal to the contact surface, by which they are pressed together. Therefore during the period in which the hypothetic boulder communicates an excess of pressure to a small area of the rock bed, the same area experiences a proportionate excess of force in a direction lying in the plane of contact. The composition of this force with the

differential force normal to the plane of contact gives a resultant parallel to the general system of oblique stresses in the surrounding ice.[29]

Moreover, the pattern of these gouges represented a "mechanical rhythm," an erosional ripple. The rupture, a "miniature earthquake," followed an accumulation of frictional strain. From this point Gilbert reasoned that "the ordinary movement of a glacier in its bed may be rhythmic" also. Because the pressure on the boulder resulted from the "resistance of the ice to the forcing of a boulder into it," the ice had enormous viscosity, a resistance to flow which increased directly with velocity. Consequently Gilbert conceived glacial flow, as Chamberlin did, as a product of interstitial melting and regelation by rigid crystalline grains.[30]

It was a tightly argued essay, a distillation of the style characteristic of his finer monographs. With a blend of analysis and description, Gilbert had constructed a chain of wrought-iron reasoning which moved from the observation of a tiny, eccentric topographic feature to the physics of glacial flow.

He did much the same with his celebrated "The Convexity of Hilltops" in 1909. The problematic topographic feature here examined was an exception to the law of divides he had enunciated in the *Henry Mountains*. In 1877 he had confessed his inability to produce a satisfactory explanation of why badland hills had rounded tops. But, like so many other riddles, this one only receded into his subconscious. Shortly before he became ill in 1909, however, he discovered the solution. He ingeniously combined an 1892 proposal by William Morris Davis that convexity resulted from gravitational creep with the accidental laboratory provided by hydraulic-mining debris. The mines had blasted enormous quantities of river gravels and redeposited them as thousands of "miniature hills." Davis' terse article had been largely ignored. But at Nevada City, California, Gilbert was able to incorporate it permanently into geomorphic thought.

In essence, GK conceived of a graded hillslope. In his 1876 notebooks, he had speculated that the badland hills were a massive body, sculptured into convex form by differential weathering. In 1909, studying mounds of homogeneous debris, he ascribed the cause instead to differential transportation. Moreover, he brought the transportation of debris by creep into close alignment with its transportation by water—both shaped a profile which conformed to the operation of gravity. "On the upper slopes, where water currents are

Gilbert's photograph, ca. 1907, of the miniature gravel hills formed during hydraulic mining. *Courtesy of the USGS Photographic Library.*

weak, soil creep dominates and profiles are convex. On lower slopes waterflow dominates and profiles are concave." Gilbert imagined that the transportation process was in equilibrium. The slope was everywhere adjusted to provide just the velocity sufficient to move a uniform layer of surface debris. "In other words," he concluded, "the normal product to degradation by creep is a profile convex upward."[31]

Gravity similarly provided the explanation for one of Gilbert's rare forays into petrology. Seeing the opportunity to direct attention toward what he regarded as "a superb field for the study of the mechanics and physics of large plutonic bodies," he attacked the problem of "Gravitational Assemblage in Granite" in 1906. The phenomenon of banding, plus reservoirs of large, dense crystals, suggested the hypothesis that granite crystals were "assembled by gravity, being either lighter or heavier than the magma from which they had crystallized." Their sheer size supported this suggestion, for as

Gilbert observed "the propelling force, differential weight, is proportional to the cube of the diameter, while the resistance of the magma is proportional to the square of the diameter."[32]

The problem of gravitational loading and transport in magma had its complement in the problem of stress unloading in solidified plutons. One of the first features of the Sierras to strike Gilbert was the topography of granite domes, which paralleled curved planar joints. The existing theories of jointing naturally tried to correlate joint fractures with the contractional hypothesis, classifying joints either by analogy to shrinkage cracks or, more commonly, by analogy to slaty cleavage formed by compression. As early as 1882 GK had found both theories wanting. Basing his opinion on observations of recent fissures in barely consolidated materials he had discovered at Bonneville, he insisted that "neither hypothesis is satisfactory, and the problem is an open one."[33]

In the case of Sierra domes, the debate hinged on whether the joints determined the dome form or whether the domes created the jointing. Le Conte, Muir, and Matthes—studying the question during his famous topographic survey of Yosemite—held the former theory: the joints had resulted from regional tectonic warping and the domes merely reflected preferential weathering along those predetermined patterns. After a thoughtful examination, supplemented by tours to rock quarries, Gilbert disagreed. The joints, he held, were a product of unloading by surface erosion and the removal of glacial ice cover; they represented rupture along lines of tensile stress. In this way, the existing forms became self-perpetuating: the more erosion, the more jointing; the more joints fracturing the landscape, the more erosion. Differential stress release thus worked in the same way as other geologic processes, such as streams, glaciers, and shore waves; that is, it simplified the contours of the landscape. But Gilbert opposed this force by another, "glacial corrasion." The actual domes of the Sierra crest line were a product, or a ratio, of the two.

With these brief essays, GK concluded a lifetime reconnaissance of glaciation. The glacial hypothesis—the Ice Age first proclaimed by Louis Agassiz—had appeared a scant three years before his birth. He literally grew up with its sweeping visions of glacial invasions, and the concept became one of the more tantalizing problems in nineteenth-century geology. He explored the characteristics of continental, tidal, valley, and cirque glaciers; he studied the complementary question of pluvial lakes in Utah, Ohio, and New York; he scrutinized the evidence for postglacial effects, notably stress release; he criticized most of the dominant theories for the origin of the glacial periods; and he reasoned the consequences of changing ice and lake flow for Niagara Falls. Yet his intellectual journey from Rochester to the High Sierras never yielded a major theory of glaciation. What appealed to him was not a glacial history but the freshness of the glacial record, not a glacial stratigraphy so much as the mechanics of glacial flow. He was curious less about the problems of the Pleistocene than about the processes of geomorphology and geophysics it exhibited: he used the recent past as a key to the present.

While his many papers and the great *Lake Bonneville* comprise a marvelous reconnaissance of the subject, they were never consolidated into a distinctly Gilbertian interpretation. For the most part, they consist of brilliant observations, thoughtful prose, and trenchant criticisms but not a unified theory. There is, however, no evidence that Gilbert desired to write anything more. What had inspired Agassiz, and what subsequently impressed those Americans who began their careers through him, was the scale and unity of the Ice Age. The search to discover the age and size of the earth thus contained a miniature quest to determine the dates and extent of the ice sheets. That perturbations were discovered in the glacial advances no more deflected that search than did the fact that the evolution of the crust involved a series of orogenic revolutions, or that the evolution of landscapes could pass through a series of partially completed cycles, or that organic evolution had exhibited periods of acceleration and retardation. These were merely ornamentations on the larger architecture of nature's progress. The unit of analysis remained the evolutionary or thermodynamic cycle.

In the early 1870s the question of Pleistocene climate arrested Gilbert's attention; he actively debated the theories of James Croll and others. By the 1890s he had abandoned the search for the origin of climatic changes, content to study their actual consequences on the earth's surface and to demonstrate that a given Pleistocene climate could generate glacials in one environment and pluvials in another. By 1900 he had suggested that the larger climate was meaningless to explain the actual, local patterns of Pleistocene effects. By 1910 he had abdicated the climatic question entirely, preferring to examine the mechanics of glacial flow and sculpture, looking at the glacier as he did the water in his flumes. By this time he was exploring glacial geomorphlogy less for its clues about climate than for the information it revealed about the physics of glacial motion.

Where Agassiz saw in the Ice Age a splendid illustration of geological history by catastrophe, where Muir saw a metaphor for God, where Chamberlin saw the orbits of an evolutionary cosmology, where King and Le Conte saw in existing glaciers a source of national pride, Gilbert saw a physiographic engine, addressing the general question of the Pleistocene through its microcosm, Niagara Falls, whose oscillating recession was intimately bound up with Ice Age history. For Agassiz, as for those who followed his example, the specter of the Ice Age intersected the syndrome of romanticism. With its strange relics, its exotic landforms, and the melancholic panorama of global destruction, it conjured up an extinct world that belonged with Newberry's carboniferous salamanders, Stephens' lost civilizations in the Yucatán, and Scott's Waverley novels. As the century learned more about the Pleistocene that simple vision became more complex, and its metaphysical splendor less dramatic, but the fundamental historical and etiological unity of the epoch was never doubted.

For GK, however, the Pleistocene, like Niagara, counteracted that syndrome. It opened up a Pandora's box of proliferating variables, interrelationships, and qualifications that all but made a mockery of geological chronologies and historical causality and that consequently deprived the concept of its romantic connotations. Niagara was but a valve in the complicated machinery of the Pleistocene; it had to be understood within the context of a network of interrelated pipes, baffles, pumps, and hydraulics. Despite the tantalizing but ultimately maddening challenge to date it, the falls seemed to defy historical organization, in the sense of history either as a causal mechanism or as a heuristic framework. The same held for the whole continental ice sheet. The single idea which had most guided Pleistocene studies, and which was epitomized in the spectacle of an ice invasion, was the vision of a global climate change. In the hands of Chamberlin, that vision was firmly consolidated into the biography of the planet; in the hands of Gilbert, it was splintered into a kaleidoscope of particularized effects which disintegrated the cosmological unity of the age.

Gilbert the Geophysicist

During these same years, Gilbert wrapped up his investigations into a second dominant field of inquiry—structural geology, including both crustal tectonics and the internal structure of the earth. As a result of new research, he published concluding thoughts on the meaning of the geothermal gradient and the significance of gravitational anomalies for the theory of isostasy. Most of his attention, however, grappled with the mechanism and landforms of earthquakes and faulting: the horizontal motion of the San Andreas fault and the vertical movements along the Wasatch.

By 1904 Gilbert had exhausted the $1,000 preliminary grant from the Carnegie Institution to investigate his plans for deep boring. Costs made the scheme quixotic, and it remains so even today. His own observations, he confessed to Joseph Barrell in 1915, "will of course never be published, and I confide them to you only because of the possibility that you may find among them some useful suggestion."[34]

Instead he turned to geomorphology. It had been habitual with Gilbert to reduce geomorphic terranes to structural terms whenever possible; it was natural enough to correlate the forms of those structures, in turn, to geophysical forces. That was, in a sense, what he had done with the mean plain. But the extrapolations possible from such data were meager. In an 1896 lecture at Johns Hopkins on the temperature of the earth's interior, he concluded half facetiously that "what we now know of the interior heat of the earth may be expressed by—???000°." Until a rational theory of matter emerged, he felt further empirical curves were of little use. Better mathematical-mechanical theories for purely physical phenomena were needed before fundamental questions about the earth could be meaningfully attacked. Given those superior theories, suitable analogies could translate them into geological environments. In 1886 GK considered the nebular hypothesis the "most probable" of the lot; by 1896, at this lecture, he despaired of nearly all such theories. In his 1915 letter to Barrell, however, he expressed a "prejudice in favor of a very hot centrosphere—so hot as to be quite adequate to maintain the gradient." He nowhere mentioned radioactive decay as a possible source of heat nor, more predictably, did he say anything about the heat source origin or about its ultimate end.[35]

The distribution of planetary heat blended well with the problem of density distributions throughout the globe. In 1914, critiquing articles by American geodesists J. F. Hayford and William Bowie, Gilbert reexamined the geologic significance of isostasy. He argued that the cause for gravitational anomalies was "an effect of heterogeneity in the nucleus and an effect of irregularity in the vertical distribution of densities within the crust." Many geologic events frequently cited as producing anomalies actually involved "a running-down process." As with his critique of progressive time, Gilbert saw instead "a perpetual initiative of geologic activity," a rhythm, which countered these tendencies. As with his analysis of surface landforms, he conceived the internal forms of the earth to be a product of processes tending to uniformity and processes creating diversity. Still attempting to correlate geologic and geodetic data, he felt it was "quite possible that underflow in a mobile layer might effect a practically perfect adjustment for differences in density above the layer, so as to bring crustal densities and crustal relief into harmony, and yet leave uncompensated the differences in density of the nucleus." But, in the end, he reluctantly conceded that "the inner earth is the inalienable playground for the imagination."[36]

While isostasy and the geothermal gradient puzzled Gilbert by their meaning regarding the structure and dynamics of the earth's interior, earthquakes and faulting brought those subjects to the earth's crust—he repeatedly tried to answer the causes of one complementary phenomenon through the other. In its broader sense, as a description of the physical texture of surface rock patterns as well as of tectonic deformations, structural geology underwrote his interpretation of landforms. Hence GK had solid reasons for studying the process of faulting. He gave a comprehensive description of the two characteristic fault types of the West by his investigation of the San Andreas fault and the Basin Range faults, particularly the Wasatch. In 1904 he prefaced this scientific campaign with a tribute to his old comrade, Clarence Dutton. Reviewing Dutton's final book, *Earthquakes in the Light of the New Seismology*, Gilbert congratulated his friend for bringing to the task "the mental equipment resulting from prolonged study of volcanism and the greater problems of the inner earth." The text consolidated seismology as it had developed after the study of the Charleston earthquake of 1886, but it also prepared seismology—and Gilbert—for the investigation of the San Francisco earthquake of 1906.[37]

The anatomy of the 1906 earthquake quickly escalated into a major scientific enterprise. Almost immediately, informal groups of geologists rallied around J. C. Branner at Stanford and A. C. Lawson at Berkeley. Within a week the governor of California appointed the California Earthquake Commission to report on the disaster. A platoon of Japanese geophysicists, architects, and engineers from the Imperial Earthquake Investigating Committee appeared to offer advice—Japan being at the time the foremost center for earthquake research. Eventually the U.S. Geological Survey fielded a team of geologists and engineers which, in conjunction with the Army Corps of Engineers, made recommendations on building materials and architectural designs to mitigate future earthquake damage. Gilbert, still smarting from previous failures—having been "tantalized by narrowly missing the great Inyo earthquake of 1872 and the Alaska earthquake of 1899"—cheerfully, even gratefully, accepted appointments to both the Geological Survey and the state commissions.[38]

Most of his work was descriptive. In every way except the scope of his personally conducted research, Gilbert's earthquake study was an ideal companion to his concomitant work in hydraulic mining. For one thing, as analyses of geologic structure and erosion, both studies had their origin in the *Henry Mountains*. For another, he did not hesitate to point out in both instances the practical value of the studies for both city planning and engineering. Finally, the earthquake provided a living demonstration of earth tectonics just as the hydraulic-mining debacle gave him a natural laboratory for fluvial geomorphology.

While several subcommittees mapped isoseismals and coseismals, and H. F. Reid labored over the mathematics and mechanics of elastic strain in the crust, GK spent most of his time tracing the active fault zone and photographically recording damages to structures. He spent the bulk of his "somewhat amateurish" survey, as he termed it, between the Golden Gate and Point Reyes. He was particularly interested in comparing the geomorphic evidence for horizontal, or strike-slip, faulting with the characteristics of high-angle, or normal, faults. But he searched for an acceptable mechanical explanation for the fault as well. He was not a seismologist, he confessed, yet he brazenly declared himself an "advocate of the principle of scientific trespass." While his 1907 official reports limited themselves to bland, meticulous descriptions, other popular addresses and articles put forth Reid's elastic rebound theory of earthquakes with exceptional clarity. As a frictional rhythm between force and resistance, the earthquake readily appealed to his imagination. In 1909, from this theoretical model, he assessed the possibility of earthquake forecasts.[39]

Gilbert conceived the frequency and intensity of earthquakes as obeying a "rhythmic principle." Like the astronomical cycles which influenced other geologic phenomena, earthquake intervals represented "in some way a system of rhythms," perhaps "composed of several independent rhythms, each beating with its own period; or they might contain imperfectly recorded rhythms, each requiring for its interpretation some of the less violent shocks not included in the destructive class." He imagined the timing as a series, combining properties of alternation and rhythm.[40]

The earthquake was a rupture and a movement of the earth's crust in response to accumulated strain within an elastic rock. The rock fractured when its shearing strength or elastic limit was exceeded. Thus the earthquake belonged in "a class of natural and artificial rhythms in which energy gradually passes into the potential

form as internal stress and strain and is thus stored until a resistance of fixed amounts is overcome, when a catastrophic discharge of energy takes place." The cause was "essentially a frictional
rhythm, dependent on the relation of certain rock strains and rock
stresses to the resistances afforded by adhesion and sliding friction."
It was irregular not only because the intervals of local starting and
stopping were unequal but because it was derived from a considerable area of the fault surface, in which "the local rhythms were neither harmonious nor synchronous." If the pattern of energy eruption
in an earthquake zone resembled that of geysers and violin bows, as
Gilbert suggested, the frictional rhythms which set up earthquake
waves resembled the hydraulic rhythms that produced bedforms in
a stream or the crescentic gouges under the glacial moulin.[41]

An especially attractive aspect of the general investigation concerned the significance of the earthquake belt for San Francisco.
GK's political conclusions regarding both earthquakes and hydraulic mining were virtually identical: the disasters—one natural and
one artificial—threatened the commerical status of San Francisco,
which was derived from its harbor and inland waterways. To insure
its natural destiny, the city would need a battery of engineering projects and sound scientific information. In the one case, this would
mean admitting the reality of quakes and planning accordingly; in
the other, it would mean continuing the suppression of hydraulic
mining at its source. In the fullest sense, Gilbert's warnings were a
moral as well as a scientific indictment.

In assessing the likelihood of accurate earthquake forecasts,
Gilbert noted that the identification of place was more important
than the prediction of time, as the former could insure more stringent building codes. He had nothing but reserved scorn for the low
"standard of commercial morality" which pursued a mindless boos-
terism. "This policy of assumed indifference," he remarked sternly,

> which is probably not shared by any other earthquake district
> in the world, has continued to the present time and is accom-
> panied by a policy of concealment. It is feared that if the
> ground of California has a reputation for instability, the flow
> of immigration will be checked, capital will go elsewhere, and
> business activity will be impaired. Under the influence of this
> fear, a scientific report on the earthquake of 1868 was
> suppressed.[42]

Gilbert was determined that that error would not be repeated,
and he concluded his 1907 assessment of the earthquake's meaning

with a perceptive understanding of the human and economic landscape. "The destiny of San Francisco," he observed,

> depends on the capacity and security of its harbor, on the
> wealth of the country behind it, and on its geographic relation
> to the commerce of the Pacific. Whatever the earthquake dan-
> ger may be, it is a thing to be dealt with on the ground by
> skillful engineering, not avoided by flight; and the proper basis
> for all protective measures is the fullest possible information
> as to the extent and character of the danger.[43]

It was a caveat he would repeat almost verbatim for another
earth wave which threatened the city—the more than one and a
quarter billion cubic yards of gravel slowly flooding out of the Sierra
Nevada. He was determined that, unlike Whitney's report of 1868,
neither his earthquake investigations nor his hydraulic-mining
studies would be suppressed.

To move from the rift valley of the San Andreas to the steep escarpment of the Wasatch mountains was not difficult. The Basin
Range was merely an old theme reinvigorated by the experience of
an actual earthquake—Gilbert only turned in his mule for a motor-
car. The Wasatch fault, like the San Andreas, threatened metro-
politan areas. Even in the 1880s Gilbert had warned residents of the
Great Basin, as he later cautioned the inhabitants of San Francisco,
about the perils of an active fault zone. In the 1910s both the tectonics of the Wasatch fault and the geomorphic criteria of its presence intrigued the old man. He was seventy-four when he returned
to the eastern border of the Basin Range for the last time. Had he
had sufficient time to survey the field with the mental resources he
still possessed, William Morris Davis suggests, the "resulting report
would have been without question Gilbert's greatest work." As it
was, even with its posthumous publication in incomplete form,
Gilbert's understated judgment was correct: his observations were
well "worth publishing."[44]

The purpose of the investigation was to answer criticisms made
of his early Basin Range interpretations, notably the hypothesis of
J. E. Spurr, announced in 1901, and that of his successor, C. R.
Keyes, in 1908. They had disputed both Gilbert's analysis of the tectonic movements in the region and his geomorphic criteria for identifying those faults. The revival of fieldwork in 1916, and its con-
tinuance in 1917, was his second effort to answer their critiques.
"An accident," GK remarked in laconic reference to the 1902 fire,
"prevented prompt publication of my observations, and the motive

for early publication was afterwards largely removed by the appearance of Davis's papers, which anticipated many of my results." There were, in fact, many other investigators of the field he had discovered during the Wheeler Survey.[45]

The function of Gilbert's monograph was to establish the legitimacy of structural and physiographic criteria for the identification of faults in the Basin Range. It was a problem in clarification and definition. GK was especially interested in the traits of Basin Range faulting, secondarily in its mechanics, and hardly at all in its history. Most of his time, therefore, was spent in reconstructing the maps lost in 1902, an exercise in description rather than analysis. The result is a masterpiece of close observation and reasoned use of technical language. But it lacks the vigor, freshness, and novelty of his great works. The lines of debate were drawn and the geologic language fixed before Gilbert began. By placing his work in that matrix of debate, his prose became rather formalized in conformity with an accepted geologic lexicon and style, and his thought became a catalog of observations answering specific questions rather than introducing a major reinterpretation. Despite its comprehensiveness, the product was rather routine.

Gilbert was operating within an established context which had hardly existed in 1871. In his major report for Wheeler, he had been able to contrast the Basin Range with the Colorado Plateau and the Basin Range structure with the Appalachians. In 1918 it was becoming difficult to define the Basin Range at all, and there was need for basic insights and data—either new or revived. Thus the monograph was not a pioneering study but an intensive defense of his youthful insights. He amassed in exacting detail the personal and collective experience of a generation, much of it spent along the Wasatch fault zone. When he published in 1875, it was an occasion for broad reconnaissances and bold speculations. By 1918 his treatment was more restrained: he examined a few specific areas in depth; he carefully brought his technical lexicon into conformity with the definitions set forth by the Geological Society of America; he debated the issue briskly but within a context of consolidation, not of exploration. *Studies of Basin-Range Structure* is an important document in the geologic understanding of the American West, but it is not a classic of geologic science. Davis notwithstanding, it is difficult to imagine how it could have been otherwise.

As a description of the geologic phenomenon of the Wasatch fault zone, the paper was excellent. "The youth and strength of the escarpment give it a special distinction as the representative of a type," Gilbert noted, adding that "it represents a high development of the type rather than the average development." With the Wasatch front as a standard, he compared its forms and processes with those elsewhere in the Basin Range. In his report for Wheeler, he had defined the Basin Range by several broad contrasts. In this final study, he worked only within the context of the Basin Range itself. Without systematic contrasts or general analogies to organize his material, he employed a different technique to describe his theory: he painstakingly refuted all the competing explanations. In each case he would assume the alternative argument and, with a string of simple declarative sentences, carry it to a virtual *reductio ad absurdum*. The result was a negative proof.[46]

To summarize his conclusion in these circumstances, it was necessary for Gilbert to stylistically assemble his prose into a periodic construction, which he did with remarkable success:

> Because the waves of intermontaine lakes do not create mountain fronts, because wind-made escarpments that truncate range sculpture are possible only under rare conditions of aridity and wind direction, because ice rivers have not occupied the intermont troughs of the Great Basin region, because the base line of an escarpment created by the streams draining a range is indendate, and because the base line of an escarpment created in a past age by a river traversing an intermont valley and afterward partly buried by alluvium is also indendate—therefore a range front that is an escarpment which truncates both structure and sculpture has presumably originated by faulting. The presumption is specifically strong if the piedmont rock is continuous with the rock of the range the escarpment was created or has been steepened by some erosive process.

By the time Gilbert finished, his hypothesis was no longer a presumption.[47]

Having proved his explanation for the Wasatch, GK compared it to other faults within the Basin Range province. His survey included a reexamination of the House Range, the Modoc fault near Klamath Falls, Oregon, and the Kern River fault in the southern Sierras. In Oregon he had the benefit of notes and observations made by Douglas Johnson, a devoted student of Davis', and the companionship of John Buwalda; near Bakersfield, he was guided by C. C. Moody of the University of California. Although they were not incorporated into the text of his report, he also had the benefit of those other regions he had visited in the course of his career which occupied the

Taken during Gilbert's 1901 assault on the enigma of Basin Range structure, this photograph of the House Range deftly illustrates the essential features of the typical mountain structure of the region. Photograph by Willard Johnson. *Courtesy of the USGS Photographic Library.*

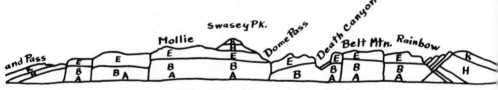

The structural composition of the House Range as redrawn from Gilbert's 1901 field notebook. One can see how, given the orientation of the two end sections, the range could be conceived as a fold, as King did, or as a collapsed arch, as Le Conte did.

other margins of the province—southern California, Arizona, New Mexico, and West Texas.

The fascination of the Basin Range problem never waned—it belonged squarely with Gilbert's indefatigable pursuit of the nature of the earth's interior and its crustal manifestations. Although he refined his initial position, he never retreated from it. In *Studies of Basin-Range Structure*, he reiterated for the final time the critique he had made against the contractional hypothesis for half a century. In a striking way this mirrored his criticism against the glacial hypothesis. "Whatever the primary force," he wrote, "and however simple, it loses simplicity in transmission through a heterogeneous crust."[48] In 1875, he had brazenly commented on the simplicity of that force; by 1918, he cautiously elaborated the complexity of its crustal evidence. To achieve for the Basin Range study that blend of description and analysis that best characterizes Gilbert's science, one must bridge the forty-three years separating the two reports.

GK's own evaluation was reserved. "I have found much occasion to regret the incompleteness of my examination," he cautioned in the text. To Barrell he frankly confided that, on the subject of the earth's interior, "these thoughts of mine are reminiscent of the time when I was fairly conversant with the literature of the greater earth problems. Had I been able to continue reading they might have been revolutionized. Your references to recent work suggest how far behind I have dropped." Consequently, the monograph ends without doing what even his essay on crescentic gouges had succeeded in— generalizing from topographic features to problems in dynamics, that is, correlating form to force. Of his own theories, Gilbert stated that he "clearly recognized their speculative nature and have lacked hope for the replacement by well-grounded theory."[49]

Yet the deficiencies are excusable. Gilbert never pretended to have completed even all those parts of the manuscript he had drafted before his death, much less to have assimilated the entire theme. The study was timely and well received, nonetheless, since the zealous William Morris Davis seized upon it as supportive of his own conclusions on the Basin Range and gladly publicized the work in order to heave Gilbert's vast prestige in favor of the validity of "physiographic evidence." That act did little violence to Gilbert's thought: to justify new criteria for faulting had been his own intention.

The study was published posthumously in 1928, the same year that Thomas Chamberlin published his last book on cosmology, shortly before his death. The two events make 1928 an appropriate conclusion to the heroic era of geologic science. Gilbert's volume made accessible all the conclusions of his first exploring days, as though a rhythm of time had finished its cycle, and Chamberlin's progressive train of thought had been pushed to its logical end with *The Two Solar Families*. But the year also brought the hesitant beginnings to the next great era of geologic science, one that would drastically supersede the larger concepts of the heroic age. It would bring about that "well-grounded theory" Gilbert had despaired of. A year earlier, in 1927, Werner Heisenberg took a giant stride in developing quantum mechanics into the new theory of matter that GK had insisted was a necessary preliminary to future speculations on planetary cosmology; a year later, in 1929, Harold Jeffreys revised his massive tome on classical geophysics, organizing its rational foundation in rheology, the general behavior of material under stress; and, in 1928, the global context of geologic tectonism underwent a fundamental debate, as the American Association of Petroleum Geologists published the proceedings of the first international symposium on the concept of continental drift.

Gilbert the Geomorphologist

Gilbert's final set of intellectual concerns dealt broadly with topics in hydraulics and fluvial geomorphology. They include three general categories of phenomena: the origin of bedforms; the recession of Niagara Falls; and the mechanisms of sediment transport by running water, both under laboratory conditions and in actual rivers.

The Medina sandstone near Niagara contains some large undulations, troughs and crests with a wave length of ten to thirty feet, which Gilbert interpreted as giant ripples. His 1898 essay on them was brief: it stands to his fluvial studies as his analysis of crescentic gouges does to his glacial work. In both cases he attempted to relate these topographic forms to the dynamics of flow. In the case of the giant ripples, he remarked, they should also "tell us something of the local physical conditions at the time of their formation." Outfitted with the theory of sand rippling recently developed by British and French physicists, he tried to reconstruct that hydraulic environment, though the "general law" was seriously flawed. When he tried to apply the "tentative law," he concluded that "the Medina ocean was agitated by storm waves sixty feet high" and that "such a result would demonstrate the association of the Medina formation with a large ocean."[50]

The reexamination of Niagara Falls had a richer background, personally and historically. In 1905, upon Gilbert's recommendation, the U.S. Geological Survey fielded the fifth official survey of Niagara, the results of which were published as "Rate of Recession of Niagara Falls." The last survey had come in 1886, when, again at Gilbert's request, Robert Woodward had plotted the crest. Now Gilbert himself returned with a small corps of assistants for a final confrontation with a topic for which his curiosity seemed insatiable. If *Studies of Basin-Range Structure* completed one cycle of research by returning him to the subject which began his exploration of the West, the Niagara Falls study completed another cycle by returning him to the topic which opened his investigation of the East. By forcing attention toward the fluctuating volumes of the Great Lakes, postglacial uplift, relic drainages, and glacial activity, the falls synthesized the themes which had preoccupied him for decades in up-state New York. His compendium of Niagara studies was as close as he came to an eastern version of *Lake Bonneville.*

It resembled the Basin Range study in other ways as well. In both there was little that was novel, except a new, more detailed set of measurements. Both existed within a confining context of pre-vious and simultaneous investigations; motifs and language were largely predetermined. Moreover, the Niagara study followed from Gilbert's laboratory work at Berkeley as the Basin Range study did from the earthquake laboratory that San Francisco had inadvertently supplied. In both cases there was a sense that what appeared to be simple was unbearably complex. Thus the purpose of Gilbert's in-quiry was modest: to identify themes and variables, to evaluate the accuracy of past survey measurements, and to program a usable lan-guage for the geologic description of the falls.

GK reserved his personal interpretation of the river for a critical review of Joseph Spencer's *The Falls of Niagara: Their Evolution and Varying Relations to the Great Lakes,* published in 1907 by the Geological Survey of Canada. Gilbert's 1908 review, reminiscent of that which he once gave Whitney, is valuable because in it he set out succinctly the fundamental analogy which guided his understand-ing of the falls. Revising Spencer's terminology, he expressed the erosional formula as follows:

> Rate of recession is proportional to the height of the falls and the discharge of the river. As the energy of the cataract (per unit time) is measured by the product of the height, into dis-charge, it is implied that the *rate of recession is proportional to the energy of the cataract.*[51]

This peculiar formulation was integral to the intellectual and aesthetic awe with which Gilbert beheld the spectacle. "I put the law into this form," he explained, "for the sake of comparing it with the experience of mechanical engineers. The cataract is a natural en-gine, and the erosion and recession correspond to what Rankine calls 'useful work' in the discussion of artificial engines." Arguing by analogy to the efficiency of heat machines, Gilbert attacked the hy-pothesis of Spencer that the efficiency (that is, recession) of the falls had been constant. On the contrary, "not merely does Spencer's sup-posed law fail to find support in engineering experience; it is contra-dicted by it." Thus, Gilbert employed the same mechanical analogy to explain the action of Niagara Falls that he had used with such effect at the Henry Mountains and Lake Bonneville. It was a model he hoped to extend generally to all landforms. "It would perhaps be more pertinent to compare the Niagara engine with other physiographic engines," he lectured, "but in general the efficiencies of such engines have not been investigated." "The solitary exception," he concluded proudly, "is that of running water regarded as a carrier of detritus, and it happens that the unpublished results of a study of this engine are in my possession." Until the mechanics of the physiographic engine were formulated, GK's contributions to the Niagara question were largely negative, criticisms of the assumptions of other estimates and the insistence on the complexity of the physical system.[52]

The scientific wonder of Niagara was that it was both complex and informative, as though the framework around a steam engine had been removed and the interior machinery were plainly visible. These same elements piqued Gilbert's sense of aesthetic wonder. "The great cataract," he marveled in 1895,

> is the embodiment of power. In every second, unceasingly, seven thousand tons of water leap from a cliff one hundred and sixty feet high, and the continuous blow they strike makes the earth tremble. It is a spectacle of great beauty. The clear, green, pouring stream, forced with growing speed against the air, parts into rhythmic jets which burst and spread till all the green is lost in a white cloud of spray, on which the rainbow floats.[53]

This rare burst of rhapsody is revealing vis-à-vis that of his contemporaries. Rather than an artistic ensemble of immense time and space, Gilbert sensed the majesty of Niagara in the spectacle of mechanical power, an apotheosis of the thermodynamic meaning of energy. Rather than an ensemble of structures organically related and progressively evolving, he saw a mechanical system naturally guided toward thermodynamic efficiency. What he found awesome, rather than merely romantic, in the great cataract resembled the amazement that hushed crowds before the Corliss engine at the Philadelphia Centennial Exposition. His enthusiasm for nature was no less than Dutton's or Powell's or Muir's or Chamberlin's—it only sprang from a subtly different sense of nature's meaning and unity.

The ideal complement to the natural cataract of power was the engineered cascade of hydraulic-mining debris. Where Niagara epitomized erosion, the Sacramento River epitomized deposition. Whereas Gilbert's interest in Niagara sprang from simple curiosity, the question of hydraulic mining constituted a major political-economic debate. But there is no doubt that the hydraulic-mining studies, both the laboratory and the field reports, are the finest expression of his later years. They justly rank with his other classics.

Like *Studies of Basin-Range Structure, The Transportation of Debris by Running Water* originated in the 1870s. Gilbert's flume experiments were a linear descendant of the *Henry Mountains*. "Thirty-five years ago," he explained,

> the writer made a study of the work of streams in shaping the face of the land. The study included a qualitative and partly deductive investigation of the laws of transportation of debris by running water, and the limitations of such methods inspired a desire for quantitative data, such as could be obtained only by experimentation with determinate conditions.

As a consequence of his study of mining debris, "instigated by the common needs of physiographic geology and hydraulic engineering," he had that opportunity.[54]

Gilbert relied on outside assistance more than usual. Besides his consultations of foreign scientific literature, he had extensive correspondence with the Coast Survey, particularly with its expert on tides, R. A. Harris. Most of the actual flume runs were conducted by E. C. Murphy. Friends like Woodward and C. E. van Orstrand criticized the manuscript. GK had enlisted Woodward's help with the problem as early as the mid 1880s, when, in preparing his Bonneville text, he asked for calculations on the hydrodynamics of shorelines. Thirty years later that partnership continued. "I ran against a mathematical snag the other day," he wrote to Arch in 1911. "Could not differentiate $\log C = n \log (S - \sigma)$ so as to get value of $d \log C / d \log S$. Tried two mathematicians—or rather engineers—I met at the Club, without result; but got an immediate return from Woodward to whom I sent it by mail. The value is $nS/S - \sigma$, and the demonstration is easy when you know how."[55]

In no aspect was Gilbert's work utterly original. Other researchers had used flumes or molded experimental rivers or tried to systematically relate bedforms to stream variables. But, like *Lake Bonneville, The Transportation of Debris* consolidated and gave vivid expression to the known facts while adding new ones; it combined the largely separate fields of hydrodynamics, hydraulics, and geomorphology. Gilbert's hydraulic lab was the fourth to be built in America, but it was the first to specifically address geologic questions, and thus it marks the origin of experimental technique in sedimentology and fluvial geomorphology. So accurate and ambitious was the project that Gilbert's figures continue to be cited today. Just as Bonneville has never been resurveyed in full, so have his flume experiments never been totally duplicated.[56]

The "primary purpose of the investigation," as GK conceived it, "was to learn the laws which control the movement of bed load, and especially to determine how the quantity of load is related to the stream's slope and discharge and to the degree of comminution of the debris." The experiments measured "capacity for hydraulic traction" against seven variables: discharge, slope, fineness of debris, depth and width (combined into a form ratio), and, to a lesser degree, mean velocity and channel curvature. "In all the range of conditions included six discharges, six widths of channel, and eight grades of sand and gravel." The possible combinations numbered almost 130, but not all were used; "the separate determinations of load and slope numbered nearly 1,200 and those of depth about 900."[57]

The experiments were conducted in four basic flumes of varying length and texture. Most of the runs came from a wooden flume 31.5 feet long and 1.96 feet wide, others from a similar trough 150 feet long. For each run a fixed discharge of water was fed into the flume, followed by a fixed amount of debris of a particular fineness. When the flume system had reached equilibrium (that is, when its slope and discharge were constant), the experiment was halted for measurements. In this way all the conditions could be held constant but one, so that the systematic effect of that one variable could be measured. Gilbert took great pains to explain his methodology and to measure the systematic and random errors endemic to his techniques.

Although this work included his most elaborate utilization of mathematics, his mathematics remained as straightforward as the channel of his flumes; he set up his equations with the same concrete inventiveness he used in designing his instruments. The reason for this similarity is simple: Gilbert viewed his formulas in physical terms, both as a graph and as having a specific physical meaning. His was a descriptive and experimental mathematics, not a deductive mathematics; he conceived his laws of stream transport in terms of mechanical rather than mathematical models. He scrutinized his equations for errors as he did the mechanics of saltation. His frustration in successfully formulating merely empirical relationships rather than a "rational theory" for stream transportation, for example, he likened to the difference between a topographic map made with hachures and one made with contour lines. Expanding

A hydraulic laboratory, ca. 1907. The largest of the four flumes which Gilbert constructed on the campus of the University of California at Berkeley. Photograph by Gilbert. *Courtesy of the USGS Photographic Library.*

the analogy, he argued that one could imagine the "capacity-slope relation as an undulating topography, in which the vertical element is C/S or $f(C)/f_1(S)$ and the horizontal elements are qualifying conditions. Formulation is a mode of representing this topography, the hills and valleys of which do not depend on the mode but are real." But he noted in exasperation that "the contour or the relief model would serve admirably if the qualifying conditions were two only, but as they number at least four, a graphic or plastic expression is possible only in space of n dimensions." In general Gilbert was best able to set up formulas so that his variables could be related directly, inversely, or by maxima or minima.[58]

GK selected his formulas so that they conformed to certain deductive properties he knew existed, to the empirical curves he drew from his flume trials, and to certain mathematical qualities, like ease of calculation and symmetry. He expected to discover a unified set of equations, both simple and predictive. But he was forced to conclude that "the development of complexity within complexity" which he repeatedly confronted "suggests that the actual nature of the relation is too involved for disentanglement by empiric methods." Yet he stubbornly clung to his hope that this "conclusion does not necessarily follow. Just as a highly complex mathematical expression may be the exact equivalent of a fairly simple expression of a different type, so a physical law may defy formulation when approached in a certain way yet yield readily when the best method of approach has been discovered." Just as his flumes failed to re-create the complex processes of a natural stream, so his mathematics failed to crystallize out that simple law he felt must operate in stream traction. In the case of both his mathematics and his flumes, his basically experimental approach may itself be an underlying cause for his disappointment.[59]

And it may well be that the flaw is even more fundamental than Gilbert imagined. The questions he asked remain elusive even to modern techniques, but they may well be insoluble by the mathematical, conceptual, and experimental techniques of Newtonian mechanics—the edifice he always worked within. That is, like riddles in Greek geometry which could not be solved within the arbitrary conditions imposed by a straightedge and a compass, so the boundary-layer problems posed by sediment transport may not be rationally solvable by the assumptions of Newtonian physics. Modern fluid mechanics, with its concepts of the boundary layer, flow net, and statistical regimes of turbulent flow, has overcome some of these difficulties, but engineers and sedimentologists have also kept Gilbert's empirical figures.

Gilbert recognized this defect, but what he said about one formula can stand for the entire study: "The modicum of physical foundation afforded . . . is believed to give it advantage over a purely mathematical expedient." It remained "a matter of faith" with him that, "if our data were so precise as to substitute definite quantitative relations for the fascicle of trends and indefinite parallelisms they have actually furnished, some way would be found leading from complexity to simplicity. I am not without hope," the old man affirmed, "that the presentation here made may suggest to the mechanist, familiar with aspects of solved problems of similar difficulty, a rational theory under which the data may advantageously be recombined."[60]

Not that the study lacked a theoretical foundation. The controlling idea returned to the fluvial models Gilbert had announced in the *Henry Mountains*. He conceived the stream as an energy system, with the profile (both longitudinal and cross-sectional) adjusted so that the stream acquired exactly the energy it needed to perform its work. In this case, work meant the transportation of its debris load. The stream derived energy because of gravity acting on channeled water; hence, its potential energy varied primarily with discharge and slope. Resisting this energy were the stream's load, its viscosity, and the fractional resistance afforded by the channel; hence, the capacity varied with velocity, amount and caliber of load, and form ratio. GK assumed that the outcome of force and resistance would be a profile of equilibrium. In experimental runs, he used this assumed steady state to decide when variables would be measured; in constructing formulas, he based his equations on the related conception of competence, that is, the measure of slope, fineness, and discharge at which the stream is capable of entraining particles. Beyond this his careful explanation of saltation, entrainment, velocity profiles, bedforms, and so on all stemmed from considerations of the acquisition and expenditure of energy. That he was predominantly concerned with slope was a consequence of his correlation of potential stream energy with declivity.[61]

Gilbert's monumental treatise was the most comprehensive work on its theme in English. In methodology and technique, in mathematical formulation, in physical explanations of stream processes, and in sheer volume of data, it was unrivaled. It summarized the known facts and, more important, posed the fundamental questions which those who followed his example would try to answer. Gilbert had established the basic concepts and definitions. He had generated a system of empirical formulas to describe the relation of capacity to variables of the stream and its load. He had put forth mechanical explanations for bedforms and modes of transportation. In the course of his text, he carefully distinguished between stream and flume traction, evaluated the influence on capacity of straight and curved channels, and assessed the significance of his flume data when applied to natural streams. Since flumes resembled the canals being erected to divert mining debris from developed areas in the Sacramento River valley, the monograph yielded some practical data for hydraulic engineering. What made the study particularly Gilbert's, however, was the fact that it occupied the interface between physics, hydraulic engineering, and geology. It was concerned not simply with the flow of water under various conditions or with fluvial erosion per se but with the interacting boundary between the two—the transportation of debris by channeled flow. It is, perhaps, in this hybridization of hydraulics and geomorphology, rather than in an exact correspondence between flumes and streams, that Gilbert's analogy of proportion best applied.

Yet it is evident that he was disappointed with his results. He dutifully enumerated his failures in both experimentation and extrapolation. He had found no experimental technique to accurately measure the tractive velocity, and there were errors associated with streamflow "rhythms" that he had been unable to eradicate. Extrapolating from his data, he conceded that all his inferred laws were "essentially inductive" and, even less happily, that "some of the laws . . . might emanate from the form of the equation and have no other basis." Contrary to the beliefs which inaugurated the study, it was doubtful whether his equations could be extended to natural streams. "If the formulae were rational, the result of an adequate mathematical treatment of the physical principles involved, the constants measured in the laboratory would be of universal application," he noted, "but the constants of an empiric formula afford no basis for extensive extrapolation." He suggested that the form of his equations might apply but not their exponents and that, "since the principles discovered in the laboratory are necessarily involved in the work of rivers," they might serve in the case of "natural streams which are geometrically similar to the laboratory streams." Citing the "natural desire" of an experimenter "to do his work over in a better way," he recommended a different approach to the problem in future studies. In the realm of experimentation, he suggested more "synthetic" modes of streams than those afforded by a flume. In terms of larger questions, "it is possible," he argued, "that the chasm between the laboratory and the river may be bridged only by an adequate theory, the work of the hydromechanist."[62]

That "adequate theory" is still elusive, and GK's greatest error was his qualified sense of failure with respect to articulating it. In one sense, *The Transportation of Debris by Running Water*, like *Studies of Basin-Range Structure*, was only a massive documentation of the abstract relationships he had announced in his youth. In another sense, as an example of method, of subjecting geologic phenomena to experimental, physical, and mathematical analysis, it stands by itself, and Gilbert ultimately did succeed in creating his model—not the rational model in hydromechanics he set out to produce but, more important, the model for its study. He may have fretted over the validity of his equations, but it is their exponent, Gilbert himself, who was confirmed in the end.

Gilbert's field report, like his laboratory treatise, was a composite. He frequently consulted experts outside geology. He benefited from earlier investigations in almost all areas of interest, including Army Corps of Engineers surveys and probes by the California Debris Commission. He had numerous intermittent assistants, especially when he was too infirm to inspect sites himself. He had regular correspondence with the Coast and Geodetic Survey, whose maps and techniques he frequently referred to, along with those of other agencies, like the Weather Bureau, the Census Bureau, the Hydrographic Office of the navy, and the Lighthouse Service. He had the least contact with other geologists, the most with engineers of various bureaus.

"The primary motive of the study," he acceded, "was economic." Under the auspices of the U.S. Geological Survey, the project was delegated to the Reclamation Service, which was intimately connected with the water resources branch. In 1906, however, Congress severed the Reclamation Service from the Geological Survey, and Gilbert was deprived of "an organization specially equipped for the study of practical engineering problems connected with the control of surface waters." The geologic data involved in the project he could handle himself; for assistance in technical and engineering questions outside geology, he necessarily turned to other bureaus. As a consequence, he again labored in a peculiar isolation, a hybrid position between physiographic geology and hydraulic engineering.[63]

The study dealt also with an uncannily complex event. Strand by strand Gilbert tirelessly unraveled the etiological tapestry of hydraulic mining. "Mining debris merged, both bodily and in its effects, with debris sent to the streams by agriculture and other industries," he explained,

> the aggravation of valley floods due to the clogging of channels by debris was inseparable from the aggravation due to the exclusion of floods from lands reclaimed for agriculture; the weakening of tidal currents at the Golden Gate by the deposition of debris in the bays is inseparable from the weakening by the reclamation of tide lands; and the attention given to these collateral subjects has not only delayed the completion of the report but has added materially to its volume.[64]

The subject amounted to nothing less than a treatise on the natural and engineered environment of the Sacramento River system, from its mangled headwaters in the Sierra Nevada to its twice-daily discharge over the tidal bar outside the Golden Gate. It demanded GK's fullest, most judicial powers, and he responded with a synthesis of the themes, techniques, and experiences culled from a lifetime of skeptical observation and controlled logic. His last completed work,

Hydraulic-Mining Debris in the Sierra Nevada, is nothing less than his intellectual autobiography.

Gilbert abandoned his usually rigid, stylized prose for one more supple and less distilled. His unit of composition was greater—no longer the sentence but the chapter. The axiomatic, deductive style characteristic of the *Henry Mountains* had climaxed in *The Transportation of Debris.* If, in the *Henry Mountains,* he had imitated formulas with prose structure, in *The Transportation of Debris* he had, because of the mathematical complexities involved, turned equations into prose equivalents. At one point he even devised a system of inflections, modeled on Greek accent marks, to designate the mathematical nature of relationships between variables. That report satisfied his desire for a formalized set of laws and processes; it freed *Hydraulic-Mining Debris* from the necessity of opening with their recapitulation.

In *Hydraulic-Mining Debris* Gilbert's passion for balance resided instead in the overall structure of the report, not its sentences, and in its tone more than its construction. His sentences were typically declarative, regularly summarized in periodic form. Inference replaced syllogism; sentences were lean but not crystalline. It was the comfortable, confident style of a master. The style was apropos, for the text proceeded not like the deductions of a mathematical proof but as a story. Almost alone among Gilbert's writings, the treatise possesses genuine narrative quality. The reason is that the contrast which underscored his analysis—the contrast between the natural and engineered regimes of the river—was broad, somewhat ill defined, and, at the time, incomplete. The differences between these two regimes were magnified by, and in part caused by, the sudden overloading of the Sacramento system with the tailings from hydraulic mining. Gilbert arrived *in medias res,* but he opened his paper with a view of the natural functions of the river. Due to commercial developments, agricultural reclamation, and mining, engineering works had drastically reconstructed the hydraulics of the Sacramento in terms of the sediment and water introduced into the system and the mechanisms by which those materials would be discharged from it. The onslaught of mining debris was not merely another instance of modification by engineering but a touchstone by which to systematically contrast the mechanisms of the natural versus the engineered regimes of the river. In following this contrast from the Sierra foothills to San Francisco Bay, GK carefully crafted his narrative until it resembled the plot of a mystery novel, even incorporating an element of suspense and featuring a surprise ending.

The quiet prose is misleading; it builds in intensity like the rising tide on a shoreline.

The organizing event was the abrupt overloading of the Sacramento River system. This mass of tailings passed through the river system in the form of a "debris wave." Yet each part of the river responded to the flood differently, and many of the natural processes of disposing of the debris had furthermore been altered by engineering devices. The conceptual formula was, thus, a familiar one: a given force acting on a system was rendered meaningful only by its manifestations in particular environments. The force in this instance was the debris wave; the local environments were the four geomorphic and hydraulic regimes of the Sacramento. Each chapter tackled a separate topic, making each a distinctive essay. Yet out of this cacophony of topics Gilbert orchestrated his analysis into a narrative of considerable power. Carefully compounding fact upon fact, inference upon inference, conclusion upon conclusion, the investigation concentrated its narrative force into a climax: the meaning of the great tidal bar outside the Golden Gate.

Yet, despite its properties as a history, the study obeyed Gilbert's conception of time. It narrated the story of a rhythm, the debris wave, from the moment it upset the natural equilibrium of the system until the time when that equilibrium was to be restored. The organizing phenomenon of the debris wave was "analogous to a flood of water in its mode of progression through a river channel. It travels as a wave, and the wave grows longer and flatter as it goes." The wave was created during the two decades of intensive hydraulic mining, concluding more or less in 1884. As it proceeded downstream, debris would accumulate until the crest of the wave passed, at which point it would begin to erode away except for a certain proportion permanently lodged in the landscape. At each locale the actual effect was a complex product of individual functions, and for each locale Gilbert had to determine the natural regime of the stream before analyzing the disequilibrium created by the sudden invasion of debris and engineering projects.[65]

He identified four broad environments through which the debris passed: the mountain valleys in which it originated; the piedmont slopes in the central valley where the Sierra streams joined the Sacramento River; the channels of the valley rivers; and the lengthy chain of bays and channels which constituted the delta of the Sacramento. At each point the laws of stream transport—and, whatever their deficiencies, he tried to apply his laboratory evidence—varied according to local conditions. The general progress of the wave had so advanced during the thirty years since the legal cessation of hydraulic mining that the flood crest was slowly leaving the piedmont for the Sacramento channel. GK predicted an additional fifty years before it would pass through the entire system, making a total cycle of eighty years.

The progress of the wave under entirely natural conditions was complex enough; compounded with the artificial factors associated with human settlement, it became inestimably more complicated. This was a new set of phenomena for Gilbert to handle. All his previous geologic studies, including his fieldwork for the 1906 earthquake, pertained entirely to the natural world, in fact, to primitive, virtually uninhabited nature. The fact that the natural processes were here resynthesized by engineering works gave him, to his benefit, the fundamental contrast that guided his critical analysis as well as, to his frustration, a vastly more complicated narrative. He had to describe the interface between the natural and human environments, and that assignment was no simpler than describing the boundary layer between running water and sand. On the one hand, variations of flood discharge, the comminution of debris, and the irregular succession of bays and channels prevented the debris wave from advancing at any straightforward rate of progression. In this respect the problem resembled that involved in determining the rate of recession for Niagara Falls. On the other hand, engineering measures also interfered with the natural processing of the debris. To divert the debris from urban and agricultural environments, a system of levees, dams, and bypass channels was being erected; at the same time, agricultural reclamation continually encroached on marsh and delta lands and demanded protection from floods of both water and debris. This altered the hydraulics of the Sacramento considerably. In his estimations, Gilbert had to take account "of variation in the activities which have contributed debris to the streams, of variation in the transporting power of streams, and of variation in certain conditions that have tended to retard or prevent the delivery of debris in the bays."[66]

"The data are vague," he lamented. Most of the story is a succession of riddles for which he must give a numerical answer. He literally weighed each clue. "Nearly all phases of the economic questions connected with the debris from hydraulic mining," he insisted, "are concerned with quantities." From fieldwork, a scrupulous reexamination of earlier estimates, and extrapolation from his laboratory work, he concluded that some 1,295,000,000 cubic yards of earth had been disturbed by mining, a "volume of earth . . . nearly eight times as great as the volume moved in making the Panama Canal." Gilbert painstakingly graphed the distribution of this debris throughout the course of the Sacramento system. With a justifiable pride, he claimed that "in the opinion of the writer, who was also the surveyor, the general accuracy is such that the grand totals are true within 10 percent, although many individual measurements have a lower precision." Revealingly, he envisioned his own estimates not as part of a progressive refinement but as a check on previous measurements.[67]

The theoretical context for comprehending the river's work was the graded stream. At the Yuba River, around Marysville, where the debris was most concentrated, he discovered a natural laboratory to test his concept. Here a dam barrier had been constructed to store the debris as one phase of the broad engineering plan enacted in 1904. By applying his law of adjusted profiles, which held that the stream adjusted its profile so that it acquired just the energy it needed to perform its work, GK showed that the dam was ephemeral both geographically and historically. As a reservoir, its up- and downstream influence extended for little more than a couple of miles. The river deposited upstream and eroded downstream in an effort to adjust its profile. Thus the life of the dam was dated. Gilbert recorded its failure in a 1906 flood and, with it, the narrowsightedness of the entire scheme to temporarily embargo or export debris via government engineering extravaganzas. He also carefully mapped the deposits left after the flood and, knowing the flood discharge from a nearby gauging station, he tried to use this information to measure the tractive velocity which had eluded him in his flumes.

Just as in his laboratory experiments Gilbert uncovered "complexity within complexity" and "variation of the rates of variation of the rates of variation," so the story of hydraulic mining was no simple melodrama but a tangled skein of characters, conflicts, and cross-purposes. He analyzed the political and economic forces operating on debris exactly as he did the forces of the natural environment. His unemotional indictment of the mining industry was matched by one against agriculture—whose own tailings, in the form of soil erosion, equaled in character and quantity those of mining by 1916. And there was yet a third disturbance: reclamation projects. One set of levees and dams served mining, another set served agriculture, but both were equally disturbing. They sought to export floodwater and flood debris from the central valley; in effect they only transferred it to San Francisco Bay. In one of his finest essays, Gilbert demonstrated the ultimate consequences.

The tidal bar outside the Golden Gate was a dynamic com-

The Yuba dam, 1906. Photograph by Gilbert. *Courtesy of the USGS Photographic Library.*

Gilbert's 1905 photograph of a hydraulic mine at Moores Flat which was drained by a tunnel to the middle Yuba River. *Courtesy of the USGS Photographic Library.*

posite whose shape and location depended on its supply of sand, the wave currents of the ocean, and the tidal (ebb) currents debouching from San Francisco Bay. "I have no knowledge of the interactions by which this profile is determined," he admitted, "but I am confident that it is in fact a profile of equilibrium, so that when any part of it is disturbed by some extraneous cause the reestablishment of equilibrium involves the reconstruction of the whole profile." By demonstrating how agricultural reclamation of Sacramento marshlands and bay filling from agricultural and industrial detritus reduced the "tidal prism"—the volume of water affected by tidal action—he showed that the cumulative effects were to steepen the tidal bar and move it closer to shore. Ultimately it would threaten the harbor entrance. Even as he wrote, dredges, at great expense, were working shoals that prior to mining and settlement had allowed ships to cross over them.[68]

GK conceived the political conflict in exactly the same way. The explosion of hydraulic, industrial mining had upset the balance between social-economic groups. Unlike many conservation documents of the day, however, Gilbert's did not naïvely pit industry against agriculture. Instead he placed both of them against "commerce"—meaning the navigability of rivers and the security of San Francisco Bay. Both industry and agriculture, outfitted with their own engineering schemes, threatened the "integrity of the harbor," which was "of higher importance to the community as a whole." Gilbert envisioned a plexus of competing economic forces, focused on the general subject of Sierra waters. He considered their "relative values," which he treated as a ratio; and he concluded that "cooperation" (translate: "equilibrium") between the interests was the only durable solution. Hydraulic mining might be tolerated under certain conditions, but the preservation of the harbor was paramount. In keeping with the requests of the miners' 1904 petition to Roosevelt, he even proposed that a "high dam" could combine limited mining with storage of Sierra waters for electrical power and agriculture. The costs were "prohibitive," however, and by 1917 the old scheme was largely irrelevant.

With its emphasis on water, its requirement that resources be used for multiple purposes and for the common good, its recommendation that the federal government supply expertise for restorative engineering projects, and its general indictment against waste, *Hydraulic-Mining Debris in the Sierra Nevada* easily belongs in the canon of progressive conservation literature. In one respect, Gilbert's monograph came much too late as a conservation document to radically change the status of hydraulic mining. Even as he traced the fracture zone of the San Andreas for the Earthquake Commission in 1906, the Miners' Association had lost all patience with his debris investigation and launched new petitions. Yet in a broader sense his text provided the first comprehensive interpretation of the Sacramento River and furnishes, even to the present day, the foundation for the management of the system. Gilbert's evaluation must be considered in the context of inland waterways schemes, amid a fierce quarrel between government agencies over who should regulate the proliferating engineering works.[69]

It is not surprising to learn that the author of *Hydraulic-Mining Debris* belonged to the Sierra Club, had been a close friend of John Wesley Powell's, and was acquainted with Van Hise, Pinchot, and most of the other leading advocates of reform or that he had spent considerable time on irrigation questions for the Wheeler and Powell surveys. What is surprising is his refusal to support agriculture against industry (unlike many of his conservationist friends, Gilbert did not come from a rural background) and the absence of belief that reform would lead to progress. GK's reforms were intended to rectify an artificial imbalance of nature, not to insure the continued political and economic progress of humankind.

Gilbert's conservation philosophy, thus, conformed to his philosophy of science. Just as his methodology differed from that of the progressive Chamberlin, so his plans for river management differed from the progressive thinking of Powell and his protégé in reform, McGee, whom Pinchot called "the scientific brains of the conservation movement." These evolutionists tended to see an intolerable scale of waste in nature's economy that nearly matched the waste in human industrial economy. Organic evolution was prodigiously extravagant with life, and physical nature tended to maximize entropy. If one could only increase the efficiency of natural processes, one could expedite the pace of evolutionary progress. With its floods and meanders, for example, and with its general degradation through time, a river maximized entropy as well as threatened human settlement. Government scientific bureaus could reduce that waste. Yet, for Gilbert, it was clear that nature could be an efficient engineer; compared to the reclamation and mining works along the Sacramento system, it was perhaps in a better position to dispose of the debris than was the California Debris Commission.

Gilbert's text differed from stock conservation productions in other respects as well. For one thing, he presented an interrelated, dynamic system, not just an inventory of water and soil resources. Unlike the evolutionists, he was able to envision this system operating as a self-regulating mechanism at any instant of time. The debris problem spanned some eighty years; the geomorphic models of landscape then prevalent involved eons of geologic time. Given to phylogenetic explanations, they were not equipped to tackle a problem interfacing with immediate questions of engineering. Gilbert was. Unlike Powell and Pinchot, he was unlikely to recklessly expand his analyses of that system into a program of national reform. As in Utah and Washington, he left politics to others. He was as unconcerned with the evolutionary future of the system as with its evolutionary past.

Less obviously perhaps, the work belongs with the literature of American nature. Just as the *Henry Mountains* and *Lake Bonneville* are a foundation for any interpretation of their regions, so is *Hydraulic-Mining Debris* for the literature of the Sierras and San Francisco Bay. There were no thrilling exploits, no melodramatic discoveries, no romantic communions with the sublime; the study's symbols were mathematical rather than literary or spiritual; its unity derived from the rigor of Newtonian mechanics rather than from transcendental oversouls or organic relationships; but throughout there is a penetrating intelligence grappling with the meaning of nature. In his suspenseful journey with the hieroglyphic riddles of mining debris, GK put his document on the Sacramento River in the category of other great scientific encounters with American rivers: the Mississippi of Humphreys and Abbot, the Colorado of Newberry and Powell, the Niagara of explorers from Champlain to James Hall to Gilbert himself. To compare Gilbert's monograph to, say, Muir's *Mountains of California* is to contrast philosophies of nature as much as literary genres. Where Muir's story takes one further and further from civilization, ultimately into a transcendental nirvana, Gilbert's takes one along the turbulent boundary between nature and civilization. The cascade of prose that Muir unleashes becomes in Gilbert's hands the deliberate, analytical sentences that belong with the disciplined banks of the Sacramento rather than with the waterfalls of the Yosemite. Yet because of this difference it was Gilbert rather than a John Muir who could make the Sacramento meaningful in a context where people were resynthesizing the landscape; his text related to the engineering of the river as well as to the grandeur of its natural regimen. Gilbert's work, as George Otis Smith—director of the USGS after 1907—memorialized, won the respect of both the geologist and the engineer: "Pure science as given to the world by Grove Karl Gilbert was useful science."[70]

The meaning of the work for Gilbert himself can be symbolized by his careful analysis of the tidal bar outside the Golden Gate. For him the bar epitomized the workings not only of the river but of nature as a whole. In the tidal bar he combined the conceptual themes of the *Henry Mountains* and *Lake Bonneville*, just as his intellectual journey down the Sacramento combined their geographic systems, the mountain stream and the bay. His investigation of the tidal bar focused on more than the waters of the Sacramento delta: it concentrated on the dominant themes and assumptions of a lifetime. In ascribing a profile of equilibrium to the tidal bar when, at the time, there was little opportunity to measure its kinetics, Gilbert was performing an act of scientific faith. His belief that the tidal bar must be in equilibrium was the counterpart to his belief that there existed a mathematical formulation for stream transport. That he had been unable to measure either was frustrating but not apostatizing. The tidal bar was the embodiment of a creed that the natural world was at its base orderly, balanced, and susceptible to mathematical-mechanical reasoning. It was a creed Gilbert never doubted.

The Cycle of Erosion: William Morris Davis

William Morris Davis possessed a faith in nature no less vigorous than that of Gilbert, but it sprang from a different metaphysic. Where Gilbert sought a rational theory based on mathematical-mechanical principles, Davis' ideal demanded a genetic classification founded on evolutionary principles. Davis envisioned himself as the synthesizer of the American school of geology. If one grants equivalent stature to Chamberlin, then in terms of his personality and conceptual apparatus Davis was right. Like his contemporaries, Henry Adams scanning the history of a civilization or Sigmund Freud plotting the psychological growth of an individual, Davis saw the natural landscape as an evolutionary system passing through progressive stages—he called that complex of events the Geographical Cycle. Combined with Chamberlin's cosmology and Norman L. Bowen's petrogenesis of igneous rocks, this meant that by 1928 evolutionism unified the geological sciences from its microcosm, the mineral, to its macrocosm, the solar system. The consequence, as William Thornbury has written, was that, "despite objections which have arisen to some of Davis's ideas, it can hardly be denied that geomorphology will probably retain his stamp longer than that of any other single person." That stamp endured for some sixty years.[71]

Davis was less an explorer than an academic synthesizer. Like most of the other major figures in geology at this time, he trained a cadre of disciples, Kirk Bryan and Douglas Johnson among them, to spread and amplify his teachings. A grandson of Lucretia Mott's, he inherited something of her moral fervor and her passion for crusading. He became himself an indefatigable propagandist and a promoter of international congresses, which could sometimes become showcases for Davisian geology; in this he resembled Powell, Chamberlin, and Le Conte. The chief intellectual nexus with his fellow geologists, however, was his unequivocal acceptance of evolutionism and a skillful application of the "naturalistic logic" that Chamberlin and Powell wielded so effectively. In 1889 he applied this logic to the new data pouring in from the western explorations. The result was a life cycle for landforms, a sequential progression of surface features beginning in youth, advancing to maturity, and declining into old age. Davis leaned heavily on Powell, Dutton, McGee, and in a different way on Gilbert. The cycle was rapidly promulgated. If Darwin had demonstrated how paleontology could be organized by the principle of natural selection, Davis showed how evolution could unify the earth fossils of geomorphology by the operation of universal denudation. His concept created a "denudation chronology" to match the "depositional chronology" of stratigraphy.

Davis did not criticize the reigning organization of geology, as history, but he attacked its refusal to accept other than paleontologic and stratigraphic criteria. He campaigned unceasingly for the legitimation of "physiographic evidence," and the reluctance of conservative geologists to accept the new data was a major reason why he, and Gilbert, often labeled themselves geographers rather than geologists. In putting that new data together into the Geographical Cycle (or Cycle of Erosion, as it was often known), Davis gave the earth sciences a badly needed innovation. The cycle envisioned a closed geologic system in which, following an initial eruption of energy in the form of uplifted land masses, there were long eons of erosion. The cycle explained what the resulting forms would look like as they systematically degenerated with time. The end result of a long, uninterrupted period would yield a nearly level erosion surface that Davis labeled a peneplain. Like the cycles of heat engines, the entropy of the landform system had increased until with the peneplain it reached a maximum state of uniformity.

The theory originally developed to explain the topography of the Appalachians, in which accordant summits (to the Davisian eye) revealed ancient, relic peneplains. The cycle could be interrupted at any stage by further uplift, a process called rejuvenation, but the remains of its former stage might still be visible—the Appalachians featured several such rejuvenated peneplains. The resulting landscape was thus as cluttered with the debris of past epochs as a Victorian drawing room with exotic bric-a-brac. As Davis put it, a little more circumspectly,

> To look upon a landscape of this kind without any recognition of the labor expended in producing it, or of the extraordinary adjustments of streams to structures and of waste to weather, is like visiting Rome in the ignorant belief that the Romans of today have no ancestors.[72]

By 1900 geology had need of new dating criteria. The production of fossils was trailing off, and there were a growing number of questions which the traditional methods could not address. Geological science turned to new evidence—structural, physical, and physiographical. With Davis the geologist was combined with the geographer, Chamberlin united the geologist with the astronomer, John Joly with the physicist. Chamberlin reconstructed geochronology on the basis of "dynamic vestiges" and diastrophism; Joly substituted radioactivity as a heat source in place of contraction; and Davis introduced the earth fossils of landforms. Yet the universal chronologies remained; the model of the heat engine persisted; and, of course, the earth continued irrevocably onward both to its ultimate human evolution and to its heat death. Davis' Cycle of Erosion was only a miniaturization of the larger cycle of earth history.[73]

From his chair at Harvard, then later from the University of Arizona and California Polytechnic, Davis elaborated the language, structure, and deductive tone of his universal cycle. It was, everyone gratefully acknowledged, well adapted to teaching. Where Gilbert sought to make his work practical by packaging it in ways useful to geologists, Davis sought to organize it for students. The system was verbal, visual, and deductive. Davis had exceptional talent as an artist, and many of his landform drawings are classics of scientific illustration. Much of his data, especially in his earlier years, he derived by pondering over geologic maps. Even his language had a direct, homey appeal. Many of his terms, as Kirk Bryan defiantly rebutted, "are essentially metaphorical. They sum up a complex idea in a verbal picture." Speaking of Davis' reliance on organic terms like "youth," "maturity," and "old age," Bryan argued that the three together "carry the simile between the life of man and the decay and wastage of original landforms. This is poetry summed in three words."[74]

Such was his artistic skill that Davis could argue with similar conviction through his diagrams. He could hypothesize an ideal landscape, then show how it would evolve by a series of drawings illustrating each stage of degradation—his pictures were theorems as much as illustrations. In much the same way, he created his own language, generating a lexicon of technical terms that resembled paradigms from a Latin grammar; one could almost decline landscape evolution through the sequence of its appropriate terminology. This verbal declension of terminology combined with the visual logic of his diagrams to create an impressive pseudomathematical structure. Together they approximated a geometric precision; the cycle spoke with visual and verbal syllogisms. Equally, Davis' elaborate terminologies—and his drawings—have the same function as the differential calculus: they grasp the fleeting instant of geological dynamics, in this case, the dynamics of evolution. As the basic equations for his cycle proliferated, so did the tables of Davis' visual and verbal derivatives. And in view of these achievements Bryan could respond with some rhetorical flourish to critics of the cycle: "Who may dissect or confine the breadth of the imagination?"[75]

The creation of the Cycle of Erosion was unquestionably one of the great acts of imagination in American geology. Davis followed up its publication with other works almost equally powerful—studies in the Basin Range, for example, and his classic work on coral reefs. Yet by 1905 there were cracks in the smooth masonry of the cycle; by 1920 major revisions were underway which continued until his death in 1934 at age eighty-four. The reason is simple. The cycle was breaking down when applied to regions which were exotic vis-à-vis the Appalachians. Or, to put it another way, the information base for geology continued to increase exponentially, as it had from the time of Werner and Hutton.

By 1905 reforms were necessary to salvage the natural laws of the science. The reconstruction of geochronological schemes by Chamberlin, Joly, and Davis was one response; Davis' cycle was another. It is easy to ridicule the mental gymnastics that the earth sciences underwent at this time, simple to accuse the crusading Davis of intellectual hypocrisy. Yet that would be wrong. What he attempted in geology had direct parallels to certain events in physics. The effort to preserve the natural laws of mechanics through the transformations of relativity theory was exactly correlative to the Davisian effort to relativize the natural laws embodied in the Geographical Cycle. Just as systems of physical laws were seen to vary according to the relative velocity of the system and its observer, so systems of evolutionary laws varied according to the climates and erosive rates of the system and to the role of the observer in modifying it. Soon there were cycles for arid lands, for glacial lands, for marine shorelines, for karstic environments, and so forth. Although each distinct cycle varied in its rates and resulting landforms, the general principle of evolution was preserved. Chamberlin and Davis would each have found the chance universe as unpalatable as had Einstein. When challenged by Walter Penck in Germany or by instances of exceptional landforms, Davis sought to incorporate them as special cases of his general laws in a way analogous to Einstein's futile search for a unified field theory. With nothing to take its place, to abandon evolutionism meant intellectual anarchy.

William Morris Davis was superbly equipped for the task. But those skills were precisely the ones that differentiated him from Gilbert. The relationship between the two men is therefore rather curious. Not only was Davis the acknowledged master of the field in which Gilbert produced some of his most notable essays, but it was Davis who, at the request of Charles Walcott, wrote the National Academy of Sciences' memoir of Gilbert. The longest memoir of its distinguished series, the biography was presented to the academy in 1922 and published in 1926. It revealed what is confirmed by other sources: the relationship between the two men was cordial, frank, and of genuine intellectual power. Yet it is doubtful that Davis understood Gilbert any more than Gilbert did Davis. Davis claimed that Gilbert instructed him "in method," and Gilbert at least once publicly defended Davis' organic terminology as "apt." Yet neither used the borrowed ideas in anything like the way their author did. The two men had more mutual respect than mutual understanding.[76]

The differences between them are wonderfully illustrated in their prose styles. Davis wrote a flexible, limpid prose; his sentences flow from one to another, arbitrarily terminated by punctuation marks. It is an adaptable, opportunistic, noncommittal style, well adjusted to an evolutionary world of perpetual change. Its elasticity is extraordinary. In his memoir of Gilbert, Davis is capable of close and extended paraphrase, without having his style lose shape, for as much as several hundred words; by changing only a few phrases and sprinkling a few semicolons and commas, he can virtually quote another source without visible distortion. It was precisely this quality which allowed him to synthesize conceptually, as well as stylistically, the work of his contemporaries.[77]

In stark contrast, Gilbert saw a world whose forms were less fluid and more distinct, a world of almost classical structure which he imitated in his prose. He presented a spatially dynamic landscape in which, for the span of time considered, the shape of the landform was preserved; Davis manufactured a historically dynamic landscape, in which landforms were continually dissolving before time.

To compare the structure of their language is to compare the poetry of Alexander Pope to that of Walt Whitman. Yet to accuse Davis of mere metaphor is to ignore the concepts that his language carried. His life cycle, after all, is no more metaphorical than Gilbert's physiographic engines; his qualitative prose should no more be faulted than that of his intellectual mentor, Darwin.

The conceptual frameworks of the two men differed from their prose in exact proportion. To the welcome onslaught of new information the versatile Davis responded by coining new terms, new concepts, even a new science, as though science, like evolution, spawned new genera and species in the course of adapting to new conditions. Gilbert, however, invented terms reluctantly, preferring to take an old word and give it new meaning. He did not find radically new landforms in the West but analogues of scenes he knew in New York, just as he did not see new principles acting on that landscape so much as analogues of old ones. Consequently, his interpretive designs have a strangely classical shape—coming upon them in an age of romantic evolutionism is like discovering one of those early railroad depots built as a Greek temple. It is equally revealing to see how both men applied the techniques of landform analysis to features far removed from the American West. Where Gilbert, on the example of Newton, puzzled over the moon, Davis, on the example of Dana and Darwin, studied coral reefs.

Their use of the graded stream concept illustrates perfectly the core conceptual, as distinguished from perceptual, difference between them. For Gilbert it meant an adjustment of energy to work—in particular, an adjustment of river slope. For Davis it meant a moment in time—that is, a stage in the history of the river. Gilbert saw a thermodynamic landscape which tried to maximize its mechanical efficiency, and whose forms were profiles of equilibrium; Davis saw a landscape increasing in entropy, whose efficiency and forms degenerated through time. Gilbert imagined an environment shaped by the maximal energy of the processes acting on it, such as floods; Davis was impressed with the cumulative effects of the slow, almost infinitesimal operation of everyday processes. Given an imagination as supple as his prose, Davis easily absorbed Gilbert's concepts into his own amoeboid theories.

He did the same thing with Gilbert's biography. Throughout the memoir, he repeatedly inserted elaborate apologies for Gilbert's failures to read the Davisian answer into particular problems. As a result, the bias in the biography is almost identical to that present in Davis' translation of Penck, and one can only conclude that the bafflement was genuine. When the pieces of the Davisian puzzle failed to fit, GK was enthusiastically portrayed as an anticipator or pathbreaker for Davis' revolutionary synthesis. Nothing less could have justified the monumental labor involved in the memoir. It was clearly an invitation for a protégé to write an equally massive memorial of Davis.

Yet none did. Perhaps his Protean talents were too slippery, or his stature too imposing. Certainly the Geographical Cycle became both. In a fantastic passage, Davis once defended himself against the charge of rigid categorization. "Certain it is that when various kinds and degrees of interruption at various stages in a cycle have been considered," he noted ingenuously, "the variety of possible combinations becomes so great that there is no difficulty whatever in matching the variety of nature. The difficulty is indeed reversed; there are not enough kinds of observed facts on the small earth in the momentary present to match the long list of deduced elements of the scheme."[78] No doubt Davis meant that his schema, like Mendeleev's periodic chart, would supply the needed categories in the search for new geologic elements. But it may be more accurate to suggest that he resembled scholastics debating the great Chain of Being, plotting categories of creatures not known to exist except by force of logic. Davis strove for comprehensiveness, the grand scheme beyond the particular fact; Gilbert for intensiveness, the essential attribute in any system or event. In his thirst for elaboration, Davis' theories acquired the restless design of a Gothic cathedral, while Gilbert's impulse to distill gave his monographs the crisp elegance of the Parthenon.

It was inevitable that the Davisian synthesis would disintegrate. The edifice rose, swollen by its talent to absorb new information, until like the vaulted arch at Beauvais it collapsed from its own unwieldy weight and aspirations. With painful irony the critiques Davis leveled against the physical geographers before him were exactly those turned against him many decades later. The system was no longer functional, causal, or holistic. During its heyday, as Kirk Bryan observed approvingly, "a host of adherents gathered facts, filled in detail, and rarely questioned his leadership."[79] But soon the master's maps were filled and overlayed until the landscape became a virtual palimpsest. To cope with new situations his students, after his example, carved and compounded the Davisian schema until their historical landscapes resembled the maps of Ptolemaic astronomy. Instead of the clean lines of the newly developed Geographical Cycle, there was a mélange of epicycles and equants overlying basic orbits in a maze that became unintelligible, even sadly comical. At times scholastic debate threatened to replace scientific discussion until the logic of Davisian methodology seemed perilously close to asking how many peneplains could sit on the head of a mountain.

At the same time, there was a shift in the attention of geomorphology. Where Davis deliberately addressed a pure, natural landscape, a new methodology emerged to grapple with the synthetic one. It related most directly to engineering. Equally there was a revolution in the metaphysics of the science, one incorporating concepts of information, systems theory, and the steady state. A new synthesis of major earth processes has occurred with the theory of plate tectonics, and modern genetics has vaporized the metaphysical and moral energy behind geology's long defense of biological evolutionism. Geology, in short, is solving different questions for the culture: the problems of a contracting or steady-state existence rather than an expanding or evolutionary one. The critique of the Davisian system is not that it is wrong so much as that it is no longer germane; it cannot address these new questions. It demanded an earth with eons of time and a broad wilderness frontier. The new earth sciences deal with neither.

This was not a situation Davis could have predicted, but it is one he could have easily understood. What allowed him to integrate the western explorations of the American school was the panorama of geologic time, as well as the conviction that what organized the landscape of time was evolution. Every feature in the human or natural landscape underwent its evolution—from words to social institutions, from mountains to scientific theories. It was this mechanism, combined with a dazzling vision of endless, unflagging, inexorable time, that powered Davis' geology. Like the ether of physics, geologic time saturated the natural world from the tiniest mud crack to the span of nebulas. Where Gilbert related the human and natural environment through the mechanical efficiencies that characterized their engineering works, Davis brought the organic and inorganic landscapes together through the awesome sublimity of geologic time. In his memoir of Gilbert, he expressed this vision in a rather poignant passage. "Would that the pencilled outlines in the little pocket diaries have been written out elsewhere more at length," he declaimed, "and yet how short would have been their endurance as the centuries roll by even had they been engraved on tablets of stone with an iron quill."[80] No one would have appreciated better than William Morris Davis the erosion which time had wrought on his own works.

A New Life

That was not an appreciation which Gilbert longed to cultivate. In his personal life, as in his geologic landscapes, he preferred to live in the present tense. When his hydraulic studies were finally published in 1917, he was seventy-four years old. Repeatedly his last monographs had been threatened by his crippling illness. Yet, with hardly more than a pause for the galley proofs, Gilbert returned to the desert mountains of the Basin Range province and prepared to begin a new life.

During his Sierra Club outings, GK had become acquainted with Alice Eastwood, a botanist with the California Academy of Sciences. Their friendship matured during his years at Berkeley, and by 1918 their affection had ripened to the point where they decided to marry. "Alice and I have been lovers for years," he wrote Arch,

> but for a long time I would not propose marriage because it seemed like asking her to give up a life that satisfied her to become the nurse of my broken health. Later I hoped to make my home with you and was willing to let Fran and you care for me when the unjust need should come. When that hope failed, I found I was unwilling to become a burden to the Merriams, and their mode of life has now so changed that my decrepitude might seriously embarrass them. At the same time my general health has so far improved that I am less ashamed to impose myself on Alice, and it looks as though for years I might be a pleasure to her instead of a burden.

While everyone else continued to call him Karl, the botanist Alice called him Grove.[81]

Following in his father's path, Arch had become a civil engineer involved with dam construction, a career that sent him to remote sites for long periods of time. His wife, Fran, died young, and he was left with a son, Palmer Grove. The grandfather, with Alice's help, proposed to raise "P. G." But Gilbert qualified the plan by remarking that he had "no personal interest in him, and do not expect to have until his mind develops. Helpless babyhood does not appeal to me at all—it may even be an annoyance." Fortunately it did appeal to Alice, and the entire arrangement pleased Gilbert. "I am anticipating so great a change for the better in my life," he explained to Arch, "that you need not condole with me. I am changing from the nomadic to the stable, from the club to the home, from friends to a wife—and you know what all that means."[82]

One of the most famous of Gilbert's photographs, this 1906 photo shows the fault trace of that year's earthquake near Olema, California. The woman is probably Alice Eastwood. *Courtesy of the USGS Photographic Library.*

He planned to celebrate his seventy-fifth birthday in California, after a layover at Emma's house in Michigan. Congratulatory letters from around the world poured in. Whitman Cross wrote that "the general recognition of his contributions to science had made him, I believe, the most highly esteemed geologist of America and one of the most widely known in the world." Emmanuel de Margerie remarked sagely,

> When so many geologists seem to be shaken by the fever of publishing, you have been among the few and wise who refrain from such hasty proceedings, always deferring the final expression of your thought till you felt it was ripe—here again, a model to follow!

E. C. Andrews congratulated him for combining the talents of the field geologist with the techniques of the "geodetical man." "When I think of the real leaders of science, the men who have lifted us along, it is Darwin, Faraday, Hutton, Lyell, Mendel, J. J. Thompson, Kelvin, Smith (William), and the other great ones . . . ," he wrote. "Among them it is my belief your name will be placed, namely, as one who could see the main principle emerging from the merge of details." The Royal Society of London made him a corresponding member.[83]

All this was heady tribute for a man who, as Merriam put it, was

> an authority in many fields, and yet one who never assumed authority; a leader in science, and yet one who never assumed leadership; neither power nor glory did he seek, but the satisfaction of contributing his share to the sum of human knowledge.

Yet Gilbert's "qualities were those that appealed to the heart as well as the mind." That double influence was not negligible. George Otis Smith observed that

> by reason of his earlier connection with two of the earlier Federal surveys, the Wheeler and Powell Surveys, and his continuous service of 39 years with the present organization, Mr. Gilbert had perhaps the largest opportunity to influence the working geologists in the Federal service, and that he so fully improved that opportunity was due to his exceptional talent for scientific leadership.

His "combination of keen mind and kind heart . . . made him a leader in scientific thought, whose share in making and fixing the standards of the United States Geological Survey cannot be overestimated." Henry Fairchild compared Gilbert to "a sort of father-advisor to the members of the Survey. Doubtless much of his thought found expression in the writings of the younger men who revered and loved him." Gilbert, Fairchild recalled fondly, never said "a harsh word of anyone."[84]

And so it went. An old friend from the Ohio Survey, J. J. Stevenson, summed up one side: "Your Lake Bonneville and Henry Mountains have more genuine thoughts in them than one finds in a cord (wood measure) of the [U.S. Geological Survey] bulletins." William Morris Davis added a second:

It would be a serious error to imagine that Gilbert's influence
on geology and geologists is to be measured chiefly by the volume of his scientific work. The manner set forth impressed
many a reader of his reports and addresses as deeply as the
results themselves. Moreover, his high personal character exerted a most beneficent influence upon all who came into contact with him, and they were many . . .

A modern critic, James Gilluly, has elaborated that point:

His significance was not revolutionary, like Gibbs, but like
Robert Welkman, our first Nobel laureate, in the general excel-
lence of his works. In his example rather than in the novelty of
his philosophy is the reason for the tremendous impact of
G. K. Gilbert upon American and world geology.[85]

Collectively the accolades testify to GK's uncanny ability to exist in isolation yet have his ideas absorbed into the mainstream of
geologic thought as he personally, despite his eremitic habits, was
accepted in scientific social circles. In the final analysis, the two
phenomena may prove inextricable: his science and his temperament strangely mirrored one another. There is no paradox to the fact
that his personality was as memorable and honored as his written
legacy—it was his example as scientist and man that endured
beyond his particular conclusions and techniques. When he wrote
that scientific method was inculcated by example, it was less that
particular essay than his whole personal career that worked to prove
it; and it is for his own example, for the inculcation of Grove Karl
Gilbert, that he is remembered. That is praise even the scrupulously
modest Gilbert would have enjoyed.

He earned a reputation in the twentieth century which would
have surprised him. In his own day his scientific temperament was
classical and conservative rather than romantic or revolutionary. He
was the antithesis of the explorers so common in his time, the scientific buccaneers brazenly carting back to civilization the plunder
they had gained from their adventures. On the contrary, Gilbert preferred assimilation to discovery; he proceeded by analogy rather
than invention. Confronted with new facts, strange landforms, and
exotic geologic structures, he did not respond by creating equally
new terms or advocating new sciences or applying revolutionary
principles. Instead he sought to explain the marvels by applying the
existing principles of the older sciences extended through analogy.

Such considerations again type Gilbert as the antithesis of the romantic artist, the Melville or the Alfred Wegener, the wasted, creative genius banished into oblivion until the future should vindicate him. Rather Gilbert was the Mark Twain, celebrated in his own time but subject to continual reinterpretations by each subsequent generation. While it may be paradoxical that our accidental geologist should be honored for qualities that appeal to the present rather than as a fallen monument of the past, that fact is not inappropriate for one who preferred the glistening pebbles of Lake Bonneville to the hoary bones of a mastodon. His concepts continue to speak in the present tense. That, too, Gilbert would have liked.

It was a perception he carried through to the end. While at Jackson his health collapsed. At Foote Hospital he spent several weeks slowly recovering. On April 7, he wrote Arch to come. He wrote Roy soon afterward. On April 22, a pragmatist to the end, he wrote David White, chief geologist of the Survey, that he had "suffered another physical slump" but boldly hoped "to be able to continue work on the report at hand—the structure of the Basin Range—and shall try to complete it." "An outside estimate for the current fiscal year," he wrote cautiously, "will be $150." And from his sickbed Gilbert labored over his own accounts. This demanded considerable effort, but he persevered until ultimately the books balanced. In truth it had taken some seven decades and the span of a continent before that final tally could be made. But what began in the Nutshell on the shores of Lake Ontario and ended at the tidal bar outside the Golden Gate at last squared.[86]

The question of what nature was and how it should behave had merged inextricably with the problem of what man was and how he should conduct his life; the problems posed by a boy's birth in 1843 proved inseparable from the riddles of the Rochester landscape. The two were solved simultaneously. Nearby Ontario became the Great Lakes of the Holocene, the igneous "lake-rocks" of the Henrys, the Pleistocene Bonneville, the buried reservoirs beneath the High Plains, the lunar maria, and San Francisco Bay. The Genesee River became the Escalante, the Colorado, the Niagara, and the Sacramento. The warped shorelines around Rochester led him to those at Erie, at Bonneville, and to the crustal undulations of the planet. The glacial topography of Rochester brought him to the tumbling moraines of Ohio and the arid shore bars etched into the Wasatch; the enigma of glacial climate brought him to the raw fjords of Alaska and the jeweled lakes of the High Sierras. He generated himself the sympathy he saw abundantly in nature and regulated himself with the same self-control, equally serene and undemanding. Now the accounts squared. Gilbert drew a small box to indicate that fact at the bottom of the page.

A few days later he packed to leave. There were more studies beckoning in the Utah desert; with Alice Eastwood, there was a new wife; with his grandson, a new family. There was a new life promised at San Francisco to replace that he had known in Rochester and Washingon. But the old life had run its course. His heart failed. He died on May 1, five days before his seventy-fifth birthday.

Arch telegraphed the news to Merriam. The stack of letters congratulating him on his birthday—an honor roll of international geology—stayed unread. On May 27 he was cremated. His ashes were brought to his family plot. There they were placed over his wife's grave.

Notes

1. In a Nutshell

1. Letter from Gilbert to William Dall, 08–01–02, in William Dall Papers, Record Unit 7073, Smithsonian Archives; I am indebted to Alan Bain, an archivist at the Smithsonian, for bringing the letter to my attention. The relevant genealogical information is summarized in William Morris Davis, *Biographical Memoir of Grove Karl Gilbert, 1843–1918*, National Academy of Sciences Biographical Memoirs 21, no. 2 (Washington, D.C.: Government Printing Office, 1926). Along with the Davis materials, there are records held by the family of Karl Gilbert Palmer; these are hereafter known as the Palmer Collection.

2. See the descriptions given by Archibald Gilbert to W. C. Mendenhall, in a letter, 10–31–18, in the Palmer Collection.

3. Letter from sister of Eliza Stanley Gilbert, 10–04–53, Palmer Collection; letter from George Barker, 12–13–53, Palmer Collection.

4. Gilbert's enthusiasm for riddles is well documented in E. C. Andrews, "Grove Karl Gilbert," Sierra Club *Bulletin* 11 (1920–23): 60–68.

5. John Merz, *A History of European Scientific Thought in the Nineteenth Century*, 2 vols. (New York: Dover, 1965 [1902]), 2: 144.

6. The story comes from Andrews, "Grove Karl Gilbert," p. 63. The problem can be solved in a number of ways. The simplest method follows. The volume of a sphere is $\frac{4}{3}\pi r^3$; of a hemisphere, $\frac{2}{3}\pi r^3$. Call the radius of the "crumb" r_1 and the radius of the hemisphere comprising the "crust" r_2. The volume of the crumb is $V_1 = \frac{2}{3}\pi r_1^3$. The volume of the crust, however, is the volume of the hemisphere of radius r_2 minus the volume of the crumb. Hence the volume of the crust is $V_2 = \frac{2}{3}\pi r_2^3 - \frac{2}{3}\pi r_1^3$. Since the volume of the crust equals that of the crumb, set the two equations equal to each other:

$$V_1 = V_2$$
$$\frac{2}{3}\pi r_1^3 = \frac{2}{3}\pi r_2^3 - \frac{2}{3}\pi r_1^3$$
$$r_1^3 = r_2^3 - r_1^3$$
$$2r_1^3 = r_2^3$$
$$r_1^3 = r_2^3/2$$
$$r_1 = r_2/\sqrt[3]{2} = .7937r_2$$

7. Davis, *Biographical Memoir*, p. 6. In the original text, Gilbert spelled "heart" as "hart." This was in keeping with his irregular adoption of the simplified spelling reforms. Unfortunately he was not consistent, so to prevent misunderstandings I have uniformly corrected his spelling—though not his punctuation—in his unpublished works so that it conforms to current orthography. For example, Gilbert replaced "height" with "hight," which only offers more confusion than clarification.

8. For his remarks on Cosmos Hall, see Gilbert, "Memoir of Edwin E. Howell," Geological Society of America Bulletin 23 (1912): 30.

9. The material on Humboldt is large. There are three usable biographies: Helmut de Terra, *Humboldt: The Life and Times of Alexander von Humboldt, 1769–1859* (New York: Knopf, 1955); Douglas Botting, *Humboldt and the Cosmos* (New York: Harper & Row, 1973); and Charlotte Kellner, *Alexander von Humboldt* (New York: Oxford University Press, 1963). Two other interesting descriptions are Erwin Ackerknecht, "Georg Forster, Alexander von Humboldt, and Ethnology," *Isis* 46 (September 1955): 83–95, and a chapter in Victor von Hagen, *South America Called Them* (New York: Knopf, 1945). The best accounts of Humboldtean influences in the United States are in William Goetzmann's *Army Exploration in the American West, 1803–1863* (New Haven: Yale University Press, 1959) and *Exploration and Empire* (New York: Knopf, 1966). I should add that Goetzmann has aptly dubbed the opening of the continental interiors "the second great age of discovery," and, of course, I am indebted to him for insights into the general intellectual problem that the discoveries created. See also his "Paradigm Lost," in Nathan Reingold, ed., *The Sciences in the American Context: New Perspectives* (Washington, D.C.: Smithsonian Press), pp. 21–34.

10. From Botting, *Humboldt and the Cosmos*, p. 76.

11. Ibid., p. 257.

12. For an assessment of Humboldt's contributions to geography, see Preston James, *All Possible Worlds: A History of Geographical Ideas* (Indianapolis: Odyssey Press, 1972).

13. A copy of the newspaper account is pasted into Gilbert's bound volumes of his publications, now stored in the U.S. Geological Survey National Center Library. Davis, *Biographical Memoir*, p. 7, quotes the article almost in its entirety.

14. Gilbert, "Notes of Investigation at Cohoes with Reference to the Circumstances of the Deposition of the Skeleton of the Mastodon," *21st Annual Report*, State Cabinet of Natural History (Albany, 1871): 137–139. The episode can be placed in the larger context of knowledge about mastodons with Robert Silverberg, *Mammoths, Mastodons, and Man* (New York: McGraw-Hill, 1970).

2. "Astride the occidental mule"

1. The story of Gilbert's hiring is told in Davis, *Biographical Memoir*, p. 11.
2. John Newberry, "Report of Progress in 1869," *Geological Survey of Ohio* (Columbus: Columbus Printing Co., 1870), p. 7. George Merrill, *The First One Hundred Years of American Geology* (New Haven: Yale University Press, 1924), summarizes the histories of nineteenth-century state surveys until 1879, when the U.S. Geological Survey was founded. A good overview of the state surveys is available in an article by Michele Aldrich in the *Dictionary of American History*, s.v. "state surveys."
3. John Newberry, "Report of Progress in 1870," *Geological Survey of Ohio* (Columbus: Nevins & Myers, 1871), p. 8.
4. Quotation is from Davis, *Biographical Memoir*, p. 11.
5. The instructions are reprinted in Newberry, "Report of Progress in 1869," p. 11. They are also pasted to the inside cover of Gilbert's Ohio field notebooks, stored in Record Group 57, National Archives. The notebooks give a fair idea of how geologic investigation was actually carried out.
6. There is no biography of Newberry or collection of Newberry papers. My information derives from J. F. Kemp, "Memorial of J. S. Newberry," *Geological Society of America Bulletin* 4 (1893): 393–406; George Merrill, "John Strong Newberry," *Dictionary of American Biography*; various references in Merrill, *The First One Hundred Years*; and, for Newberry's labors on the western surveys, Goetzmann, *Army Exploration*.
7. For the Ives Expedition, see Goetzmann, *Army Exploration*, chap. 10.
8. Newberry's remarks are in Lt. Joseph Ives, *Report Upon the Colorado River of the West* (Washington, D.C.: Government Printing Office, 1861), pt. 2: 45. For an evaluation of the impact of Newberry's comments, particularly as received in Europe, consult Richard Chorley et al., *The History of the Study of Landforms; or, The Development of Geomorphology* (London: Methuen, 1964), vol. 1.
9. Joseph Le Conte, "Memorial of James Dwight Dana," *Geological Society of America Bulletin* 7 (1895): 463; John Newberry, "Geological Relations of Ohio," *Report of the Geological Survey of Ohio*, 4 vols. (Columbus: Nevins & Myers, 1873), 1: 51. For an influential essay in which Newberry put forth the terms of reconciliation between the new science and the old religion, see his "Presidential Address," *Proceedings of the American Association for the Advancement of Science* 16 (1869): 1–16.
10. Newberry, "Report of Progress in 1870," pp. 32–33.
11. For the fight with Whittlesey, see Merrill, *The First One Hundred Years*, pp. 451–453. Newberry quotations are from his "Report of Progress in 1870," p. 14.
12. For the Newberry letter to Hayes and its effects, see Thomas Manning,

Government in Science: The U.S. Geological Survey (Lexington: University of Kentucky Press, 1967), p. 58.

13. Merrill, *The First One Hundred Years*, p. 452; Gilbert Field Notebook 3365, Record Group 57, National Archives.

14. See Gilbert, "Surface Geology of the Maumee Valley," *Report of the Geological Survey of Ohio* 1: 541.

15. Ibid., pp. 553, 540. As a result of these observations, H. E. Gregory credited Gilbert with the first unequivocal recognition of terminal moraines in the United States. See Gregory, "Steps of Progress in the Interpretation of Landforms," *American Journal of Science* 46 (July 1918): 104–132.

16. Gilbert, "Some Facts in Regard to the Surface Geology of the Maumee Valley," *Transactions of the New York Lyceum of Natural History* (New York, 1871), 1: 178.

17. Gilbert, "On Certain Glacial and Postglacial Phenomena of the Maumee Valley," *American Journal of Science* 3d ser. 1 (1871): 344.

18. Gilbert, "Surface Geology of the Maumee Valley," p. 552.

19. The surveys have been the subject of considerable research. There are two general books: Richard Bartlett, *Great Surveys of the American West* (Norman: University of Oklahoma Press, 1962), and Goetzmann, *Exploration and Empire*. The King Survey is also discussed in Thurman Wilkens, *Clarence King* (New York: Macmillan, 1958); the Powell Survey in William Culp Darrah, *Powell of the Colorado* (Princeton: Princeton University Press, 1950), and Wallace Stegner, *Beyond the Hundredth Meridian* (Boston: Houghton Mifflin, 1953). My material on the Wheeler Survey comes from the progress reports issued annually by it, Wheeler's 1889 report, and the unpublished records housed in Record Group 77, National Archives, supplemented by William Goetzmann, "The Wheeler Surveys and the Decline of Army Exploration in the West," in Robert Ferris, ed., *The American West: An Appraisal* (Santa Fe: Museum of New Mexico Press, 1963).

20. Capt. George M. Wheeler, "Geographical Report," *U.S. Geographical Survey West of the 100th Meridian* (Washington, D.C.: Government Printing Office, 1889), p. 31. This was Wheeler's "final report" and will be referred to as such hereafter.

21. For the quoted phrase, see Wheeler, "Final Report," p. 46.

22. Gilbert, May 9, 1871, Field Notebook 3372.

23. Gilbert, May 25, July 6, June 7, 1871, Field Notebook 3372.

24. Gilbert, July 30, 1871, Field Notebook 3373.

25. Gilbert, July 31, 1871, Field Notebook 3373.

26. For the stories printed about Wheeler in Death Valley, see Bartlett, *Great Surveys*, pp. 342–344. Quotations as follows: Wheeler, "Final Report," p. 45; Frederick Loring, "Into the Valley of Death," *Appleton's Journal* 6 (1871): 575; Gilbert, July 30, 1871, Field Notebook 3373.

27. Gilbert, July 31, August 1, 1871, Field Notebook 3373; Gilbert, August 27, August 31, 1871, Field Notebook 3374.

28. Gilbert, August 25, August 19, 1871, Field Notebook 3374; Gilbert,

September 16, 1871, Field Notebook 3375. My account follows closely
that preserved in Gilbert's field notebooks. Wheeler wrote a fuller narrative in his 1889 report but, since the boats were frequently separated, I
have adhered to GK's version.

29. Gilbert, September 21, October 1, 1871, Field Notebook 3375.
30. Gilbert, September 1, September 22, September 23, September 30,
    1871, Field Notebook 3375.
31. Gilbert, October 4, October 7, 1871, Field Notebook 3375.
32. Gilbert, October 1, 1871, Field Notebook 3375.
33. Gilbert, October 13, 1871, Field Notebook 3375.
34. Gilbert, October 15, October 16, 1871, Field Notebook 3375.
35. Gilbert, October 18, 1871, Field Notebook 3375.
36. Gilbert, October 18, 1871, Field Notebook 3375.
37. Gilbert, October 20, 1871, Field Notebook 3375.
38. Wheeler, "Final Report," p. 170; Lt. George M. Wheeler, "Preliminary
    Report Concerning Explorations and Surveys, Principally in Nevada
    and Arizona" (Washington, D.C.: Government Printing Office, 1872),
    p. 19.
39. Gilbert, October 28, November 1, 1871, Field Notebook 3375; Wheeler,
    "Final Report," p. 45; Gilbert, December 6, 1871, Field Notebook 3375.
40. Wheeler, "Preliminary Report," p. 60.
41. Gilbert, October 17, 1872, Field Notebook 3381; Gilbert, October 23,
    1872, Field Notebook 3382.
42. Gilbert, November 13, October 24, 1872, Field Notebook 3382.
43. Lt. George M. Wheeler, "Annual Report of the Geographical Explora-
    tions and Surveys West of the 100th Meridian," *Appendix FF of Annual
    Report of the Chief of Engineers for 1874* (Washington, D.C.: Govern-
    ment Printing Office, 1874), p. 103. Henry Henshaw, a naturalist with
    Wheeler from 1872 to 1874, later an associate with Powell in the Bu-
    reau of Ethnology, and finally an ornithologist with the Biological Sur-
    vey, offered a more sympathetic interpretation: "It is but fair to say that
    no one recognized the limitations of the scientific man's opportunities
    more clearly than Wheeler, but he himself was greatly hampered by of-
    ficial rulings and requirements from higher up, and no doubt the con-
    duct of the survey he inaugurated would have been very different had
    he been free to follow his own ideas." Letter of 11–11–18, Palmer
    Collection.
44. Gilbert, October 6, 1873, Field Notebook 3386.
45. Gilbert, August 20, 1873, Field Notebook 3385.
46. Letter from Wheeler to Gilbert, 06–07–74, "Letters Sent," Wheeler Sur-
    vey, Record Group 77, National Archives; Gilbert, July 17, 1872, Field
    Notebook 3379.
47. Gilbert, "Preliminary Geology Report," in Lt. George M. Wheeler,
    "Progress Report Upon the Geographical and Geological Survey West of
    the 100th Meridian" (Washington, D.C.: Government Printing Office,
    1873), p. 48.
48. Gilbert, "Preliminary Geology Report," p. 48.

49. The priority for naming the Colorado Plateau, like all questions of priority, is muddled. Wheeler insisted that the name was his but confused his claim by giving two different dates for his invention of the term. Gilbert bestowed the honor upon Newberry, and perhaps it belongs there. Gilbert himself was notoriously indifferent to claims of priority, yet it was his use of the term—especially in contrast to the Basin Range—that gave the expression general currency.

50. Gilbert, "Geology," *Report of the U.S. Geographical and Geological Survey West of the 100th Meridian* (Washington, D.C.: Government Printing Office, 1875), 3: 41, 61, 62.

51. The Geikie quote is from Gordon Davies, *Earth in Decay* (New York: Neal Watson, 1969), p. 352. In his memoir of Gilbert, Davis recounts several incidents illustrating European stubbornness on the subject of normal faulting in the Basin Range. Even after Gilbert, in 1891, personally escorted some members of the International Geographic Congress to fresh fault scarps along the Wasatch, they refused to accept his explanation—until, after several days of repeated demonstrations, they finally capitulated. The process of conversion was even slower when personal examination was not possible. The great range in Europe, of course, was the Alps, which, like the Appalachians, provided vivid testimony of orogeny by crustal shortening.

52. Gilbert, "Geology," p. 118; Clarence Dutton, *Report on the Geology of the High Plateaus of Utah*, U.S. Geological and Geographical Survey of the Rocky Mountain Region (Washington, D.C.: Government Printing Office, 1880).

53. Gilbert, "Geology," p. 559.

54. Gilbert, "Preliminary Geology Report," p. 49.

55. Gilbert, "Geology," p. 106. The consequences of deforestation were not original with Gilbert or Powell. In his second volume of *Cosmos*, Humboldt remarked: "By felling the trees which cover the tops and sides of mountains, men in every climate prepare at once two calamities for future generations: want of fuel and scarcity of water" (p. 154). The message was repeated by George Perkins Marsh prior to the *Arid Lands* report; see his *Man and Nature* (New York: Charles Scribner, 1864).

56. Dutton quoted by Gordon Craig and Mary Rabbit, "Sir Archibald Geikie, Letters from American Geologists," *Geotimes* 14, no. 5 (May–June 1969): 21.

57. Henshaw letter, 11–11–18, Palmer Collection.

58. Ibid.

59. In Gilbert, "Preliminary Geology Report," p. 92; Gilbert, "Geology," p. 33.

60. Gilbert letter to F. B. Meek, 01–04–74, "Letters Received," Meek Papers, Smithsonian Archives. In full, the note reads: "I discovered a few days ago that you had immortalized me on a bryozoan [*Olenus gilberti*]. Thank you for remembering me. It is the first time I have been thus honored. Mr. Howell, who is still with me, wishes to be remembered." And so he was, in *Olenus howelli*. I am indebted to Clifford

Nelson for directing me to these letters. A carbon of Gilbert's typewritten insert (dated 02–26–76) is still present in his copy of the Wheeler report, now stored in the U.S. Geological Survey National Center Library, and the full text is reproduced in Davis, *Biographical Memoir*, pp. 25–26.

61. Wheeler to Gilbert, 10–23–74, "Letters Sent," Wheeler Survey. Wheeler himself was clearly caught in the middle on this issue between Gilbert, who had not yet published anything for the survey, except progress reports, and Humphreys, who had no desire to see things published outside of it. Eventually the articles appeared shortly before the debut of the official report, as well as somewhat earlier at a meeting of the American Association for the Advancement of Science in the form of addresses which were eventually printed in its proceedings.

62. See the entry on Wheeler in the *Dictionary of American Biography*.

63. Lt. George M. Wheeler, "Report on the Third International Geographic Congress and Exhibition," *House Executive Document* 270, 48th Cong. 2d sess. (Washington, D.C.: Government Printing Office, 1881), pp. 157, 55.

64. Consider this sample of Wheeler's inimitable prose: "However sanguine the inhabitants of the various western sections may be who desire to see growing up in their midst a chain of settlements occupied by industrious agriculturalists, still it must be remembered that, with all the assistance that may be rendered by the general Government, States, and Territories, or by the total water supply provided by nature, that too much must not be expected in return for their enthusiastic attempts to reclaim its waste places" ("Preliminary Report," p. 30). In other words, there wasn't enough water. So, although his comments on western irrigation problems predate Powell's by many years, it was Powell who captured the public spotlight with his brisk prose and sharp, simple distinctions patiently outlined.

65. Wheeler, "Final Report," p. 65.

66. Gilbert to Powell, 09–01–78, "Letters Received," Powell Survey, Record Group 57, National Archives. Wheeler scornfully remarked of Powell's maps that, "so far as can be judged from the topographic maps, the base map for topography was insignificant; at least very few, if any, latitudes have been published" ("Report on the Third International Geographic Congress," p. 493).

67. Townsend Committee, "Report on the Geographical and Geological Surveys West of the Mississippi," *House Report* 612, 43d Cong., 2d sess. (Washington, D.C.: Government Printing Office, 1874), pp. 17–18.

68. Wheeler to Gilbert, 06–07–74, "Letters Sent," Wheeler Survey.

69. Gilbert to Powell, 11–27–74, "Letters Received," Powell Survey.

3. The Major Years

1. Thompson to Powell, 05–25–73, "Letters Received," Powell Survey; Stegner, *Beyond the Hundredth Meridian*, p. 127; Thompson to Powell, 03–11–72, "Letters Received," Powell Survey. Other informa-

tion on these early surveys can be found in A. H. Thompson, "Diary of A. H. Thompson, 1871–1875," *Utah Historical Quarterly* 7, nos. 1–3 (1939); Dale Morgan, ed., "The Exploration of the Colorado River and the High Plateaus in 1871–1872," *Utah Historical Quarterly* 16, no. 17 (1947–48); and Frederick Delenbaugh, *A Canyon Voyage* (New Haven: Yale University Press, 1962).

2. William Dall to F. B. Meek, 01–27–75, "Letters Received," Museum of Natural History Library, MSS 31. Actually Gilbert had written Meek earlier in the month asking permission to use his library—as he had when they collaborated on the Ohio and Wheeler surveys. Gilbert's labors these days are summarized in a letter to Powell, 03–19–75, "Letters Received," Powell Survey.

3. Gilbert, August 20, 1875, Field Notebook 3393.

4. Gilbert, August 23, August 24, 1875, Field Notebook 3393.

5. Gilbert, August 22, 1875, Field Notebook 3393.

6. Gilbert, August 30, 1875, Field Notebook 3394.

7. Gilbert, August 11, 1876, Field Notebook 3396; Gilbert, September 9, 1876, Field Notebook 3396; Gilbert to Powell, 09–24–75, "Letters Received," Powell Survey. Gilbert also wrote revealingly that "I can make a structure model of nearly the whole country I have seen and I want to do it." He did, but the important thing is that from very early he conceived of the problems in regional and three-dimensional terms, not local or topographical ones.

8. Gilbert, *Report on the Geology of the Henry Mountains* (Washington, D.C.: Government Printing Office, 1877), p. vii. The field notebooks demonstrate that this is precisely the sort of investigation he conducted.

9. Gilbert, October 11, 1876, Field Notebook 3398; Gilbert, October 12, 1876, Field Notebook 3398. The pages also include several diagrams very similar to those which appeared in the final report. Most of the illustrations in the report, in fact, derive from Gilbert's field notes; his pencil sketches are much better done than the printed lithographs suggest.

10. Gilbert, October 13, October 15, 1876, Field Notebook 3398; Gilbert, October 20, October 21, 1876, Field Notebook 3399.

11. Gilbert, October 27, 1876, Field Notebook 3399.

12. Gilbert, *Henry Mountains*, pp. vii–viii.

13. Gilbert, November 13, 1876, Field Notebook 3400; Dutton to Powell, 09–17–76, "Letters Received," Powell Survey.

14. Gilbert, November 2, 1876, Field Notebook 3400.

15. Gilbert to Powell, 07–18–77, "Letters Received," Powell Survey; Gilbert, August 25, 1877, Field Notebook 3404.

16. Willard Johnson to W J McGee, 04–11–85, W J McGee Papers, Library of Congress.

17. Gilbert to Powell, 08–26–78, "Letters Received," Powell Survey.

18. Gilbert, October 27, 1878, Field Notebook 3409; Gilbert to Powell,

09–01–78, "Letters Received," Powell Survey; Gilbert, October 8, 1878, Field Notebook 3409.

19. Gilbert to Powell, 10–18–78 and 11–01–78, "Letters Received," Powell Survey.

20. Gilbert to Powell, 10–05–78, "Letters Received," Powell Survey.

21. Clarence Dutton, *Mount Taylor and the Zuñi Plateau*, U.S. Geological Survey *Annual Report* 6 (Washington, D.C.: Government Printing Office, 1885), p. 113; Dutton, *Tertiary History of the Grand Canyon District*, U.S. Geological Survey Monograph 2 (Washington, D.C.: Government Printing Office, 1882), p. 143.

22. Davis, *Biographical Memoir*, p. 71. Information on Powell and the Civil War experience is found in George Frederickson, *The Inner Civil War* (New York: Harper & Row, 1965).

23. Powell, "Biographical Notice of Archibald Robertson Marvine," Philosophical Society of Washington *Bulletin* 3 (1878): 53. The address was delivered on June 3, 1876. In eulogizing Marvine, Powell was outlining his own understanding of geology—and praising Marvine for conforming to it. The lecture is especially valuable because it was given in the same year that *The Geology of the Uinta Mountains* was published. It represents, in a sense, Powell's final thoughts on geology before leaving it—except in an administrative capacity—for anthropology. Other quotations as follows: Powell, "Biographical Notice of Archibald Marvine," p. 54; ibid., p. 57; Gilbert, "John Wesley Powell," *Proceedings of the Washington Academy of Sciences* 5 (1903): 116; Goetzmann, *Exploration and Empire*, p. 56. In his eulogy, Gilbert emphasized Powell's role as dramatist when he wrote: "It was his belief that a scientific fact needed no argument, but only statement."

24. Powell, "Biographical Notice of Archibald Marvine," pp. 58–59.

25. Powell, *The Exploration of the Colorado River of the West*, (Washington, D.C.: Government Printing Office, 1875), p. 203; Philosophical Society of Washington *Bulletin* 2 (1876): 79; Goetzmann, *Exploration and Empire*, p. 562.

26. Powell, "Biographical Notice of Archibald Marvine," p. 58. This remarkable passage may be worth quoting in full: "Some of our geological literature could be burned and no harm done. O that a pope would rise and a holy catholic church of geologists—a pope with will to issue a bull for the burning of all geological literature unsanctified by geological meaning. Then there would remain the writings of those inspired with the knowledge that a mountain has structure, that every hill has an appointed place and every river runs in a channel foreordained by earth's evolution, and Marvine's work would be a book of genesis in the bible of the geological priesthood."

27. For an indirect appraisal of Powell's impact on conservation, consult Samuel Hays, *Conservation and the Gospel of Efficiency* (New York: Atheneum, 1974).

28. Gilbert, "John Wesley Powell," *Proceedings of the Washington Acad-*

emy of Sciences, p. 113; Gilbert to Powell, 09–24–75, "Letters Received," Powell Survey.
29. Bailey Willis, *A Yanqui in Patagonia* (Palo Alto: Stanford University Press, 1947), p. 33.
30. Gilbert, "John Wesley Powell," *Science,* n.s. 61 (1902): 567.
31. Gilbert, "John Wesley Powell," *Proceedings of the Washington Academy of Sciences,* p. 115; Powell, *Exploration of the Colorado River and Its Canyons* (New York: Dover, 1961 [1895]), p. 397; Gilbert, "On the Uses of the Canyons of the Colorado for Weighing the Earth," Philosophical Society of Washington *Bulletin* 2 (1873): 88.
32. For biographical information on Dutton, see Chester Longwell, "Clarence Edward Dutton," National Academy of Sciences Biographical Memoirs 32 (1958): 136–146; J. S. Diller, "Major Clarence Edward Dutton," *Bulletin of the Seismological Society of America* 1, no. 4 (1913): 137–142; Diller, "Major Clarence Edward Dutton," Geological Society of America *Bulletin* 24 (1913): 10–18; and George Becker, "Obituary of Clarence Dutton," *American Journal of Science,* 4th ser. 33, no. 196 (1912): 387–388. Dutton has been the subject of one dissertation—Wallace Stegner, "Clarence Edward Dutton: An Appraisal" (University of Iowa, 1935), and one master's report, Stephen Pyne, "Dutton's Point: An Intellectual History of the Grand Canyon" (University of Texas at Austin, 1974).
33. Dutton, *Mount Taylor and the Zuñi Plateau,* p. 122.
34. Samuel Emmons to Becker, 06–16–82, "General Correspondence," George F. Becker Papers, Library of Congress.
35. Charles Hunt, "Geology and Geography of the Henry Mountains Region, Utah," U.S. Geological Survey Professional Paper 228 (Washington, D.C.: Government Printing Office, 1953), p. 3. Hunt also includes a detailed chronology of Gilbert's tour through the region.
36. Gilbert, *Henry Mountains,* pp. 90–91.
37. Ibid., p. 99.
38. Ibid., p. 115.
39. Gilbert, "Geology," p. 108.
40. See J. Hoover Mackin, "The Concept of the Graded River," Geological Society of American *Bulletin* 59 (1948): 471, 507. A valuable criticism of Gilbert's ideas on grade vis-à-vis those of his contemporaries is contained in Chorley et al., *History of the Study of Landforms,* 1: 609–610. A modern statement of the problem can be formulated as follows:

$$nSFP = kL^aD^b/Q^c$$

where S = slope ("grade" in Gilbert's terms), F = channel geometry, P = stream pattern, L = amount of sediment load, D = average diameter of bed load, and Q = water discharge. Equation from Marie Morisawa, *Streams: Their Dynamics and Morphology* (New York: McGraw-Hill, 1968), p. 150.
41. There is a commendable summary of the reception of the *Henry Moun-*

tains in Davis, *Biographical Memoir,* pp. 81–83, 90–95. Other criticisms can be found in Hunt, "Geology and Geography of the Henry Mountains Region," and Arvid Johnson, *Physical Processes in Geology* (San Francisco: Freeman & Co., 1970).

42. Dana's opinion—and his alternate explanation—is well discussed in Davis, *Biographical Memoir,* pp. 92–95. The contentious article challenging Gilbert's priority was by A. C. Peale, "On a Peculiar Type of Eruptive Mountains in Colorado," *Bulletin of the United States Geological and Geographical Survey of the Territories* 3 (1877): 551–564.

43. Holmes' "Random Records," now housed in the National Portrait Gallery, documents his disgruntlement about the priority problem. It should be remarked that Holmes did not blame Gilbert directly for the issue but, rather, G. P. Merrill, who credited Gilbert with the discovery in his history of geology. Of Gilbert himself, Holmes wrote a tribute which reads like the Boy Scout credo: "My personal relations with Gilbert in Washington extended over many years, and I am glad of this opportunity of recalling his many rare qualities. He was approachable, unassuming, considerate, patient, helpful, generous, companionable, cheery, appreciative, candid, consistent, well poised, wise, judicial, noble—an ideal man exercising in many ways a helpful influence upon all who came within his circle" (letter from Holmes to W. C. Mendenhall, 11–12–18, in Holmes Papers, Smithsonian Archives).

44. Gilbert, "Geology," p. 564.

45. Gilbert, "The Work of the International Geological Congress," *American Journal of Science,* 3d ser. 34 (1887): 437.

46. My biographical information on Rankine derives from Sir James Henderson, "Macquorn Rankine, An Oration" (Glasgow, 1932); Yehuda Elkana, *The Discovery of the Conservation of Energy* (London: Hutchinson Educational, 1974); and Merz, *A History of European Scientific Thought,* vol. 2. Quotation from Rankine, *A Manual of Applied Mechanics,* 9th ed. (London, 1877), p. 1.

47. Rankine, *Manual of Applied Mechanics,* p. 1.

48. Gilbert, *Henry Mountains,* p. 90.

49. Gilbert, "The Origin of Hypotheses, Illustrated by the Discussion of a Topographic Problem," *Science,* n.s. 3, no. 53 (1896): 10, 13.

50. The Philadelphia story is more elaborately narrated in Davis, *Biographical Memoir,* p. 201.

51. Henshaw letter, Palmer Collection.

52. Davis, *Biographical Memoir,* p. 118.

53. Ibid., pp. 9–10.

54. W. C. Mendenhall, "Memorial of Grove Karl Gilbert," Geological Society of America *Bulletin* 31 (1920): 42.

55. In C. Hart Merriam, "Grove Karl Gilbert, the Man," Sierra Club *Bulletin* 10, no. 4 (1919): 393–394.

56. De Margerie to Gilbert, letter in Palmer Collection.

4. A Great Engine of Research

1. Goetzmann, *Exploration and Empire*, pp. 458, 466.
2. Clarence King, U.S. Geological Survey *Annual Report* 1 (1880), pp. 76, 5. For information on the early days of the Survey, see Manning, *Government in Science*; Wilkens, *Clarence King*; Stegner, *Beyond the Hundredth Meridian*; and Goetzmann, *Exploration and Empire*.
3. See "Ascent on Mount Tyndall," in Clarence King, *Mountaineering in the Sierra Nevada* (Lincoln: University of Nebraska Press, 1970 [1872]). For some interesting evaluations of King by his contemporaries, see the *Clarence King Memoirs* (New York: Open Court Publishing Co., 1902).
4. Gilbert, U.S. Geological Survey *Annual Report* 1 (1880), pp. 26, 24, 25.
5. Ibid., p. 25.
6. Gilbert to King, 12–28–79, 11–20–79, "Letters Received," U.S. Geological Survey.
7. Gilbert, U.S. Geological Survey *Annual Report* 2 (1881), pp. 11, 15.
8. Gilbert to King, 07–22–80, 11–16–81, "Letters Received," U.S. Geological Survey.
9. Gilbert to Powell, 04–11–81, "Letters Received," U.S. Geological Survey.
10. The transformation in the Survey wrought by Powell is well described by Manning, *Government in Science*, and Goetzmann, *Exploration and Empire*.
11. Dutton quote from Craig and Rabbit, "Sir Archibald Geikie," p. 21.
12. Gilbert, U.S. Geological Survey *Annual Report* 3 (1882), p. 16.
13. Davis, *Biographical Memoir*, p. 120. The same comments can be found in the Henshaw letter, Palmer Collection.
14. Gilbert, U.S. Geological Survey *Annual Report* 4 (1884), p. 32.
15. Gilbert to Powell, 03–30–84, "Letters Received," U.S. Geological Survey.
16. Davis, *Biographical Memoir*, p. 142; Bailey Willis, *Geologic Structures* (New York: McGraw-Hill, 1923), dedication. See also Frank B. Taylor, "Correlation of Warren Beaches with Moraines in Southeastern Michigan," *American Geologist* 18 (1896): 233; and Taylor and Frank Leverett, *The Pleistocene of Indiana and Michigan and the History of the Great Lakes*, U.S. Geological Survey Monograph 53 (Washington, D.C.: Government Printing Office, 1915).
17. In Craig and Rabbit, "Sir Archibald Geikie," p. 21.
18. Davis, *Biographical Memoir*, p. 181.
19. In ibid., p. 162.
20. Ibid., pp. 162–163.
21. Ibid., p. 165.
22. In Craig and Rabbit, "Sir Archibald Geikie," p. 21.
23. Gilbert, "The Work of the International Geological Congress," pp. 434, 440.
24. Ibid., p. 451.
25. Ibid., pp. 438, 439, 437.

26. Davis, *Biographical Memoir,* p. 170.
27. Ibid., p. 169.
28. Ibid., p. 192.
29. Ibid., p. 176. The same story is related in Manning, *Government in Science,* pp. 212–213.
30. Davis, *Biographical Memoir,* p. 171.
31. Powell to Gilbert, 03–24–94, "Letters Sent," U.S. Geological Survey.
32. Biographical material on Becker is available in George Merrill, *Biographical Memoir of George Ferdinand Becker,* National Academy of Sciences Biographical Memoirs 21 (Washington, D.C.: Government Printing Office, 1920): 1–13, and in Arthur Day, "Memorial of George Ferdinand Becker," Geological Society of America *Bulletin* 31 (1919): 14–19.
33. King to Becker telegram, 03–17–81, "General Correspondence," George F. Becker Papers.
34. Becker to Hague, 04–19–89, and Emmons to Becker, 04–29–82, "General Correspondence," George F. Becker Papers.
35. For information on Barus and Woodward, see R. B. Lindsay, *Biographical Memoir of Carl Barus,* National Academy of Sciences Biographical Memoirs 22, no. 9 (Washington, D.C.: Government Printing Office, 1941): 171–192; and F. E. Wright, *Biographical Memoir of Robert Simpson Woodward,* National Academy of Sciences Biographical Memoirs 19, no. 1 (Washington, D.C.: Government Printing Office, 1937): 1–15.
36. Day, "Memorial of George Ferdinand Becker," p. 14.
37. In Merrill, *Biographical Memoir,* p. 2.
38. Becker to G. H. Darwin, 06–11–02, and Becker to Barus, 07–15–00, "General Correspondence," George F. Becker Papers.
39. Becker to Benson, 01–18–97, "General Correspondence," George F. Becker Papers.
40. Osmond Fisher, *Physics of the Earth's Crust* (Cambridge, Eng., 1881), p. 266.
41. Gilbert, "Rhythms and Geologic Time," *Science,* n.s. 11, no. 287 (1900): 1007.
42. A good survey of the numerical estimates offered in the debate can be found in Joe Burchfield, *Lord Kelvin and the Age of the Earth* (New York: Academic Press, 1975).
43. Gilbert, "Sun's Radiation and Geologic Climate," *Science,* n.s. 1, no. 16 (1883): 458.
44. The address was part of a memorial program sponsored by the Geological Society of Washington. A copy of the program is available in the Palmer Collection, but the actual minutes are evidently lost. I could find no record of Woodward's speech in the Geological Society of Washington proceedings or in the Woodward Papers at the Carnegie Institution. It is interesting to note, however, that Woodward's original degree was in civil engineering, thus creating another bond between him, Gilbert, and Rankine.
45. Gilbert, "Review of Geikie's *Geology,*" *Nature* 27, no. 689 (1883): 237.

46. Gilbert, *Lake Bonneville*, U.S. Geological Survey Monograph 1 (Washington, D.C.: Government Printing Office, 1890), pp. 1–2.
47. Ibid., p. 2.
48. Gilbert, "Review of Geikie's *Geology*," p. 239.
49. Gilbert, *Lake Bonneville*, p. 25.
50. Ibid.
51. Ibid., pp. 33–34.
52. Gilbert to Becker, 02–20–89, "General Correspondence," George F. Becker Papers.
53. Dutton's term "isostasy" comes from his famous paper, "On Some of the Greater Problems of Physical Geology," Philosophical Society of Washington *Bulletin* 11 (1889): 51–64. For a general survey of studies contemporary to Gilbert's on the subject of postglacial uplift, see Richard Flint, *Glacial Geology and the Pleistocene Period* (New York: Wiley, 1947). A good explanation of isostasy is contained in Reginald Daly, *The Strength and Structure of the Earth's Crust* (New York: Prentice-Hall, 1949).
54. Gilbert, *Lake Bonneville*, p. 380.
55. Gilbert, "Circular Rainbow Seen from a Hill-Top," *Nature* 29 (1884): 452.
56. Davis, *Biographical Memoir*, p. 134.
57. Ibid., p. 107. This passage is paraphrased many times through the course of the memoir.
58. Gilbert, "Report of the Division of the Great Basin," U.S. Geological Survey *Annual Report* 1 (1880), p. 24.
59. Gilbert, "History of the Niagara River," Commissioners of State Reservation at Niagara *Annual Report* (1889): 60–84.
60. Gilbert, "Review of Whitney's 'Climatic Changes,'" *Science* 1 (1883): 195.
61. Gilbert, "Changes in the Level of the Great Lakes," *Forum* 5 (1888): 418, 428.
62. Gilbert, "Rhythms and Geologic Time," pp. 1011, 1007.
63. For the bare bones of this episode, see Davis, *Biographical Memoir*, p. 206.
64. Gilbert, "Geology," *Appleton's Cyclopedia* (1898), p. 722.
65. Gilbert, "Ripple-marks," *Science*, n.s. 3, no. 60 (1884): 376.
66. Davis, *Biographical Memoir*, p. 183.
67. Ibid., p. 185.
68. The address was subsequently published; see Gilbert, "The Moon's Face: A Study of the Origin of Its Features," Philosophical Society of Washington *Bulletin* 12 (1893): 242. For the standard interpretation of the moon in Gilbert's day, see Edmund Neison, *The Moon and the Configuration of Its Surface* (London, 1876).
69. Gilbert, "The Moon's Face," pp. 278, 279.
70. Ibid., p. 279.
71. All of my information for this period of Gilbert's personal life comes from Davis, *Biographical Memoir*.

72. Gilbert's involvement in these societies is well documented in their published proceedings. Gilbert himself, in an official capacity as secretary, twice compiled statistics on the development of the two major organizations: the Philosophical Society of Washington and its successor, the Washington Academy of Sciences. See "Statistics of the Philosophical Society from Its Foundation (abstract)," Philosophical Society of Washington *Bulletin* 10 (1887): 29–37; and "First Annual Report of the Secretary," *Proceedings of the Washington Academy of Sciences* 1 (1899): 1–14. Gilbert was also the moving spirit in the 1893 fission of the geological branch from the Philosophical Society into a separate organization, later affiliated with the Washington Academy. This was the Geological Society of Washington; for a summary of this organization, see Frank Whitmore, "A Thousand Nights' Entertainment," *Geotimes* 20, no. 7 (July 1975): 14–15.

73. For two slightly different versions of the mess, see Davis, *Biographical Memoir*, and Manning, *Government in Science*.

74. A printed copy of Gilbert's presidential address is available in William Holmes' "Random Records," 1: 157.

75. Davis, *Biographical Memoir*, p. 123.

76. Ibid., p. 179.

5. Grade

1. Davis, *Biographical Memoir*, pp. 183, 205.

2. The map problems were especially disheartening since the original map had been done by Willard Johnson. Johnson later redeemed himself by turning his, and Gilbert's, experiences on the plains into a monograph, *The High Plains and Their Utilization*, which provided the scientific foundation for Walter Prescott Webb's *The Great Plains*.

3. Davis, *Biographical Memoir*, p. 206.

4. Gilbert, October 5, 1896, Field Notebook 3463.

5. The books are listed in Gilbert's Field Notebook 3465, after the 11–25–95 entry. The map reference is from Field Notebook 3471.

6. Dutton to Holmes, 1899, "Random Records," 8: 91–94.

7. Holmes to Mrs. Holmes, 1899, "Random Records," 2: 96.

8. The best account of the Harriman Expedition is the official version in the Harriman Alaska Series published by the Smithsonian Institution. Other versions by individual participants are abundant. In addition to Gilbert's field notes, I examined C. Hart Merriam's journals, stored at the Library of Congress, and the version given in Linnie Marsh Wolf, *Son of the Wilderness* (New York: Knopf, 1945).

9. The sequence of topics, along with references to the original papers, is thoroughly summarized in Gilbert's *Studies of Basin-Range Structure*, U.S. Geological Survey Professional Paper 153 (Washington, D.C.: Government Printing Office, 1928), pp. 264–266.

10. Davis, *Biographical Memoir*, pp. 240, 241.

11. Gilbert to Arch Gilbert, 08–03–01, Coates Collection.

12. A summary of the project is contained in Gilbert, "Plans for Obtaining

Subterranean Temperatures," Carnegie Institution of Washington *Yearbook No. 3, 1904* (Washington, D.C.: 1905), pp. 259–267. The institution took shape in 1902.

13. Gilbert, "The Underground Water of the Arkansas Valley in Eastern Colorado," U.S. Geological Survey Annual Report 17 (1896), pt. 2, p. 593.

14. Gilbert, "Tepee Buttes," Geological Society of America *Bulletin* 6 (1895): 341; Gilbert, "Lake Basins Created by Wind Erosion," *Journal of Geology* 3 (1895): 47–49; Gilbert, "Laccolites in Southeastern Colorado," *Journal of Geology* 4 (1896): 816–825.

15. Gilbert, "A Report on a Geological Examination of Some Coast and Geodetic Survey Gravity Stations," U.S. Coast and Geodetic Survey *Report for 1894*, 2 pts. (1895), pt. 2, p. 51; Gilbert, "Notes on the Gravity Determinations Reported by Mr. G. R. Putnam," Philosophical Society of Washington *Bulletin* 13 (1900): 73. The ideas of a mean plain and rock-feet are developed in this latter paper.

16. Gilbert, "A Report on a Geological Examination of Some Coast and Geodetic Survey Gravity Stations," p. 73; Gilbert, "Strength of the Earth's Crust (abstract)," Geological Society of America *Bulletin* 1 (1889): 23–25.

17. Gilbert, "A Report on a Geological Examination of Some Coast and Geodetic Survey Gravity Stations," p. 74.

18. Gilbert, *Glaciers and Glaciation*, Harriman Alaska Series (Washington, D.C.: Smithsonian Institution, 1910), vol. 3.

19. Ibid., p. 109.

20. Ibid., pp. 123, 130.

21. Ibid., pp. 203, 219.

22. Ibid., p. 221. See also an interesting footnote on p. 204 in which Gilbert disputes arguments by Becker and physical geographer N. S. Shaler—later dean of Harvard's Laurence Scientific School—on the "two theoretic limits to the law that abrasion increases with pressure." Gilbert argued that large glaciers could erode; Becker and Shaler that, because of temperature and elastic limits, there was a ceiling on the size of a glacier that could erode.

23. Ibid., p. 205.

24. Gilbert and Albert Perry Brigham, *An Introduction to Physical Geography* (New York, 1902), p. vi. The text went through three editions—1902, 1904, and 1906.

25. Gilbert, "Review of Archibald Geikie's 'Geological Sketches at Home and Abroad,' and 'Textbook of Geology,'" *Nature* 27 (1883): 237.

26. Gilbert, "Special Processes of Research," *American Journal of Science*, 3d ser. 33 (1887): 452.

27. Gilbert, "On the Inculcation of Scientific Method by Example, With an Illustration Drawn from the Quaternary Geology of Utah," *American Journal of Science*, 3d ser. 31 (1886): 284–285, 286.

28. Ibid., pp. 285–286; Davis, *Biographical Memoir*, p. 244.

29. Gilbert, "The Inculcation of Scientific Method," pp. 285–287; Gilbert, "The Origin of Hypotheses," p. 6.
30. Gilbert, "The Inculcation of Scientific Method," p. 287.
31. Ibid., p. 288.
32. Davis, *Biographical Memoir*, p. 248. The episode occurred during the 1904 Geological Society of America meeting in Saint Louis. Emmons was the critic, responding to a paper by George Louderback on some peculiar features of the Basin Range.
33. Davis, *Biographical Memoir*, p. 1.
34. Ibid., p. 269.
35. There is a wealth of material on Chamberlin—in fact, perhaps an over-abundance. The basic chronological history is available in compressed form in the *Dictionary of American Biography* and the *Dictionary of Scientific Biography*. More interpretive memoirs are R. T. Chamberlin, *Biographical Memoir of Thomas Chrowder Chamberlin, 1843–1928,* National Academy of Sciences Biographical Memoirs 15 (Washington, D.C.: Government Printing Office, 1934): 307–407; Bailey Willis, "Thomas Chrowder Chamberlin," Smithsonian Institution *Annual Report* (1929), pp. 585–593; and the entire issue of vol. 37, no. 4, of the *Journal of Geology* (May–June 1929), which contains essays on nearly every aspect of Chamberlin's thought by former associates. I am also obliged to Michele Aldrich, who combed the Chamberlin Papers for correspondence pertaining to Gilbert. Among the most interesting items are letters showing a split between Gilbert and Chamberlin over the proper organization of the Geological Survey. Chamberlin wanted more local autonomy for workers in the Glacial Division.
36. For the impact of the Civil War on Chamberlin, see C. K. Leith, "Chamberlin's Work in Wisconsin," *Journal of Geology* 37, no. 4 (May–June 1929): 289.
37. T. C. Chamberlin, "The Methods of the Earth Sciences," *Popular Science Monthly* 66 (1904–05): 73.
38. See T. C. Chamberlin, "The Development of the Planetesimal Hypothesis," *Science*, n.s. 30, no. 775 (1904): 642. The actual date for the articulation of the planetesimal hypothesis is difficult to establish, since it evolved over a period of years.
39. T. C. Chamberlin, "Studies for Students," *Journal of Geology* 5 (1897): 837–838. The preliminary version of this essay was published as "The Method of Multiple Working Hypotheses," *Science*, o.s. 5 (1890): 92–96.
40. T. C. Chamberlin, *The Origin of the Earth* (Chicago: 1916), p. 39.
41. T. C. Chamberlin, "Diastrophism as the Ultimate Basis of Correlation," *Journal of Geology* 17 (1909): 692–693.
42. T. C. Chamberlin and R. D. Salisbury, *Geology*, 3 vols. (New York: Holt, 1906), 1: iii, 3: 543.
43. Chamberlin, "Studies for Students," pp. 839, 847.
44. Ibid., p. 844.

45. Ibid., pp. 847–848.
46. T. C. Chamberlin, "Grove Karl Gilbert," *Journal of Geology* 26, no. 4 (July–August 1918): 376.
47. Willis, "Thomas Chrowder Chamberlin," p. 585.

6. The Inculcation of Grove Karl Gilbert

  1. Gilbert quote from Davis, *Biographical Memoir*, p. 295.
  2. The story of Gilbert's residency with the Merriams is best recorded in Merriam's home journals. For information on Merriam himself, consult Keir Sterling, *The Last of the Great Naturalists: The Life of C. Hart Merriam* (New York: Arno Press, 1974).
  3. Gilbert to Arch Gilbert, 05–13–11, Coates Collection.
  4. Davis, *Biographical Memoir*, p. 263.
  5. Merriam Journals, Box 10 (August 2, 1903), Library of Congress. Other information on these expeditions is available in Gilbert's field notebooks. The account is also given in Davis, *Biographical Memoir*, and Merriam, "Grove Karl Gilbert, the Man," pp. 391–396.
  6. Charles Walcott, U.S. Geological Survey *Annual Report* 25 (1905), pp. 44–45.
  7. For Gilbert's contributions to the Lawson book, see Francis Vaughn, *Andrew C. Lawson* (Glendale, Calif.: Arthur Clarke Publishing Co., 1974), p. 81.
  8. A political history of the hydraulic-mining debate is available in Robert Kelley, *Gold vs. Grain: The Hydraulic Mining Controversy in California's Sacramento Valley, A Chapter in the Decline of the Concept of Laissez Faire* (Glendale, Calif.: Arthur Clarke Publishing Co., 1959). The best history of its scientific study is the introduction to Gilbert's treatise, *Hydraulic-Mining Debris in the Sierra Nevada*, U.S. Geological Survey Professional Paper 105 (Washington, D.C.: Government Printing Office, 1917).
  9. Kelley, *Gold vs. Grain*, pp. 45, 52.
 10. Davis, *Biographical Memoir*, p. 263.
 11. Gilbert, "The Investigation of the California Earthquake of 1906," in David Starr Jordan, ed., *The California Earthquake of 1906* (San Francisco, 1907), p. 213; and Gilbert, April 20, 1906, Field Notebook 3501.
 12. Davis, *Biographical Memoir*, p. 256.
 13. Letter of E. C. Andrews, Sierra Club *Bulletin* 10, no. 4 (1919): 438.
 14. Andrews, "Grove Karl Gilbert," p. 66.
 15. Ibid., p. 67.
 16. Merriam, "Grove Karl Gilbert, the Man," p. 395; Gilbert to Arch Gilbert, 02–18–12, Coates Collection.
 17. Davis, *Biographical Memoir*, p. 264; Gilbert to Arch Gilbert, 04–04–09, Coates Collection.
 18. Davis, *Biographical Memoir*, p. 265.
 19. Ibid., pp. 277, 269, 277. A note pertaining to Gilbert's handling of the Walker Prize is contained in the Palmer Collection. The story of the Denison Library episode is referred to in a letter by Frank Carney

tion.

(04–28–10) in the Palmer Collection. For the delivery of his medals, see Gilbert to Arch Gilbert, 05–21–11, Coates Collection.

20. Davis, *Biographical Memoir*, p. 278; Gilbert to Arch Gilbert, 07–16–11, 07–30–11, 10–01–11, Coates Collection; for the critique on Murphy, see Gilbert to Arch Gilbert, 04–30–11, Coates Collection.

21. The correspondence with Willis and Barrell is preserved in the U.S. Geological Survey Field Records in Denver. One of the letters contains an interesting statement of geological creed by Willis.

22. Davis, *Biographical Memoir*, p. 269. For his opinion of Peirce, see Gilbert to Arch Gilbert, 07–28–12, Coates Collection.

23. Davis, *Biographical Memoir*, p. 269; Gilbert to Arch Gilbert, 01–26–13, Coates Collection.

24. Davis, *Biographical Memoir*, p. 269.

25. Gilbert to Arch Gilbert, 03–22–15, Coates Collection.

26. Davis, *Biographical Memoir*, p. 297.

27. Gilbert to Arch Gilbert, 04–30–17, Palmer Collection.

28. Vaughn, *Andrew C. Lawson*, pp. 309–310.

29. Gilbert, "Crescentic Gouges on Glaciated Surfaces," Geological Society of America *Bulletin* 17 (1906): 303–316.

30. Ibid., pp. 312–313.

31. Gilbert, "The Convexity of Hilltops," *Journal of Geology* 17 (1909): 346.

32. Gilbert, "Gravitational Assemblage in Granite," Geological Society of America *Bulletin* 17 (1906): 322.

33. Gilbert, "On the Origin of Jointed Structure," *American Journal of Science*, 3d ser. 24 (1882): 53. This article should be combined with one other to understand GK's general assessment of the topic: "Postglacial Joints," *American Journal of Science*, 3d ser. 23 (1882): 25–27.

34. Gilbert to Barrell, 12–13–15, U.S. Geological Survey Field Records.

35. Gilbert, "The Temperature of the Earth's Interior," 1896, unpublished manuscript in the U.S. Geological Survey Field Records.

36. Gilbert, *The Interpretation of Anomalies of Gravity*, U.S. Geological Survey Professional Paper 85 (Washington, D.C.: Government Printing Office, 1914), p. 35.

37. Gilbert, "Review of 'New Seismology,'" *Science*, n.s. 20, no. 250 (1904): 837.

38. Gilbert, "The Investigation of the California Earthquake of 1906," p. 215.

39. To learn the evolution of Gilbert's thoughts on earthquakes, contrast two of his articles: "A Theory of the Earthquakes of the Great Basin, with a Practical Application," *American Journal of Science*, 3d ser. 27 (1884): 49–53, and "Earthquake Forecasts," *Science*, n.s. 29, no. 734 (1909): 121–138.

40. Gilbert, "Earthquake Forecasts," p. 128; Gilbert, "The Earthquake as a Natural Phenomenon," U.S. Geological Survey *Bulletin* 324 (Washington, D.C.: Government Printing Office, 1907), p. 11.

41. Gilbert, "Earthquake Forecasts," p. 125.

42. Ibid., pp. 136, 135.
43. Gilbert, "The Investigation of the California Earthquake of 1906," p. 256.
44. Davis, *Biographical Memoir*, p. 298; Gilbert to David White, 04–18–18, Palmer Collection.
45. Gilbert, *Studies of Basin-Range Structure*, p. 9.
46. Ibid., p. 68.
47. Ibid., p. 47.
48. Ibid., p. 21.
49. Gilbert to Barrell, 12–13–15, U.S. Geological Survey Field Records.
50. Gilbert, "Ripple-marks and Cross-Bedding," Geological Society of America *Bulletin* 10 (1898): 137, 139.
51. Gilbert, "Evolution of Niagara Falls," *Science*, n.s. 28 (1908): 149.
52. Ibid., pp. 149, 150.
53. Gilbert, "Niagara Falls and Their History," *National Geographic Monograph* 1 (1895): 203.
54. Gilbert, *The Transportation of Debris by Running Water*, U.S. Geological Survey Professional Paper 86 (Washington, D.C.: Government Printing Office, 1914), p. 9.
55. Gilbert to Arch Gilbert, 11–12–11, Coates Collection.
56. For the general scientific context of Gilbert's fluvial studies, consult Hunter Rouse and Simon Ince, *History of Hydraulics* (Ann Arbor: University of Michigan Press, 1957), especially pp. 224–225. Some of Gilbert's correspondence and computations are preserved in Field Notes 3529 (Box 38), Record Group 57, National Archives. This box also contains the forty-three-page typescript from the notes Gilbert dictated to Mendenhall.
57. Gilbert, *The Transportation of Debris*, p. 17.
58. Ibid., p. 121.
59. Ibid., p. 109.
60. Ibid., pp. 129, 190.
61. The emphasis on verticality—manifested in his flume studies as slope—is a continual motif in nearly all of Gilbert's writings. Waterfalls, graded streams, normal faulting, isostasy—all accentuate vertical forces, at a time when emphasis was more commonly put on horizontal, compressive forces.
62. Gilbert, *The Transportation of Debris*, pp. 190, 236, 240.
63. Gilbert, *Hydraulic-Mining Debris*, pp. 7, 13.
64. Ibid., p. 7.
65. Ibid., p. 31.
66. Ibid., p. 36.
67. Ibid., pp. 38, 43, 67.
68. Ibid., p. 97.
69. For a good exposition on the relationship between engineering and conservation in this period, consult Hays, *Conservation and the Gospel of Efficiency*, especially chap. 2. Also recommended is the introduction to

Donald Worster, ed., *American Environmentalism: The Formative Period, 1860–1915* (New York: Wiley & Sons, 1973).

70. George Otis Smith, "Grove Karl Gilbert," U.S. Geological Survey *Annual Report* 39 (Washington, D.C.: Government Printing Office, 1918): 11.

71. William Thornbury, *Principles of Geomorphology* (New York: Wiley, 1954), pp. 10–11. The standard biography of Davis is Chorley et al., *History of the Study of Landforms*, vol. 2. The authors tend to approach biography as a species of documentary history. The best compendium of Davis' writings is William Morris Davis, *Geographical Essays* (New York: Dover, 1954 [1909]). A concise appraisal of Davis is contained in Ronald Flemal, "The Attack on the Davisian System," *Journal of Geologic Education* 19, no. 1 (1969): 3–13.

72. Davis, *Geographical Essays*, p. 268.

73. See Joseph Le Conte, "A Theory of the Formation of the Great Features of the Earth's Surface," *American Journal of Science*, 3d ser. 4, no. 23 (1872): 472.

74. Kirk Bryan, "Physiography," in Geological Society of America, *Geology, 1888–1938* (New York, 1939), pp. 3–4.

75. Ibid., p. 4.

76. Davis dedicated his "Structure of the Triassic Formation of the Connecticut Valley" to "G.K. Gilbert—as what there is of good in the method of discussion herein followed was suggested largely by his 'Inculcation of Scientific Method by Example.'" The dedication is inked by Davis onto a reprint of the article which is now in the Palmer Collection. Gilbert's defense of Davisian terminology can be found in "Style in Scientific Composition," *Science*, n.s. 21, no. 523 (1905): 29. To quote Gilbert further: "The stream valley resembles the human being in that from an early stage it evolves normally through a definite sequence of stages; and in most other respects the two differ." Furthermore, "in my judgment there are few groups of terms which serve better than does this group the purpose of concisely expressing an idea." It should be noted, however, that this period of his career contained Gilbert's most frequent use of Davisian terms.

77. Compare, for example, the account of house renting as preserved in the Henshaw letter, Palmer Collection, with the paraphrase used by Davis, *Biographical Memoir*, p. 68.

78. Davis, *Geographical Essays*, p. 286.

79. Bryan, "Physiography," p. 4.

80. Davis, *Biographical Memoir*, p. 10.

81. Gilbert to Arch Gilbert, 03–25–18, Palmer Collection.

82. Ibid.

83. Whitman Cross to Gilbert, 06–01–18, Palmer Collection; de Margerie to Gilbert, 04–29–18, Palmer Collection; Andrews to Gilbert, 08–02–18, Palmer Collection.

84. Merriam, "Grove Karl Gilbert, the Man," p. 396; Smith, "Grove Karl

Gilbert," p. 11; Henry Fairchild, "Grove Karl Gilbert," *Proceedings of the Rochester Academy of Sciences* 5 (1919): 258.

85. Stevenson to Gilbert, 10–12–12, U.S. Geological Survey Field Records; Davis, *Biographical Memoir*, p. 291; James Gilluly, "The Scientific Philosophy of G. K. Gilbert," in Claude Albritton, ed., *The Fabric of Geology* (Palo Alto: Freeman & Co., 1963), p. 11.

86. Davis, *Biographical Memoir*, p. 302; Gilbert to David White, 04–22–18, Palmer Collection and U.S Geological Survey Field Records.

Sources

William Morris Davis' massive memoir of Gilbert is still the largest single repository of basic Gilbert materials. Unfortunately, Davis did not have future generations of scholars in mind when he wrote it; he considered that his biography would be the last. In addition to the collections gathered by W. C. Mendenhall, Davis took considerable pains to gather his own materials prior to writing the memoir, but he took few pains to identify or save his sources. As a result, the Gilbert papers he collected have scattered, and it is almost impossible to separate many of Davis' original passages from those he quotes by paraphrase. Whenever I had recovered the original document and compared it to Davis' quotation or paraphrase, however, I discovered that Davis had quoted accurately. Hence, I had few misgivings about using Davis as a source for Gilbert anecdotes. Some of the correspondence relating to the development of the Davis memoir is housed in the archives of the National Academy of Sciences.

Of the papers which Davis gathered, those I have been able to recover belong in three groups. One cache is in the hands of the family of Gilbert's grandson, Karl Gilbert Palmer. I have referred to these documents as the Palmer Collection. They include primarily papers dealing with the very early and the very late years of GK's life, with some items on his tour of duty with the Wheeler Survey. The second cache belongs to Donald Coates, and I have referred to it as the Coates Collection. It includes more than two hundred letters, mostly to Gilbert's son, Arch, spanning about a five-year period, 1910 through 1914. Arch had originally collected the correspondence for Davis, and some of his typed explanations of events mentioned in the letters are still paper-clipped to them. A third source of Gilbert manuscripts is contained in the U.S. Geological Survey Field Records at the Denver, Colorado, office. The papers include letters between Gilbert and both Joseph Barrell and Bailey Willis. There are also unpublished manuscripts on ripple marks and on the geothermal gradient as well as Gilbert's draft of *Studies of Basin-Range Structure.*

Other primary sources are Gilbert's field notebooks and his bound set of publications. He apparently kept two sets of notebooks—one for the field and one for home. Davis refers to the home books, but disparagingly, since he found them lean and elliptical. I have not been able to locate them at all. The field notebooks, however, total some 239 and are stored in Record Group 57 of the National Archives. The accession numbers are 935–1140 and 3363–3585; there is an index available in the archives. I found the notebooks for 1871 and 1876 the most interesting—the first for its narrative value and the second for its revelations on the Henry Mountains problem. Gilbert's Henry Mountain notes have been transcribed by Arvid Johnson and David Pollard and published in photocopy format by the School of Earth Sciences, Stanford University. Box 38 (no. 3529) contains Gilbert's basic correspondence and laboratory notes for the hydraulic studies. Gilbert's published works, minus *Studies of Basin-Range Structure,* are collected into thirteen bound volumes now stored in the basement of the U.S. Geological Survey National Center Library at Reston, Virginia. The collection is valuable for preserving some articles which, like his 1867 newspaper article on the Cohoes mastodon, would be difficult to locate otherwise. A complete bibliography of Gilbert publications is contained in the biographical memoir by W. C. Mendenhall. Finally, some one thousand books out of Gilbert's personal library were donated to Denison University in Granville, Ohio, in 1905 and remain there. Since the volumes are not separately indexed, it is difficult to recover the collection for purposes of, say, reading marginal notes and the like. This kind of inspection was possible for only a handful of volumes. Many of the books were in foreign languages and were probably the result of exchanges with European geologists. Gilbert's most common reference books he clearly kept for his own use in California.

A considerable portion of Gilbert's career is recorded in the archives of, and published reports of, the organizations he served under. Record Group 77, Records of the Office of the Chief of Engineers, National Archives, contains the Wheeler Survey materials. Record Group 57, Records of the United States Geological Survey, contains both the Powell Survey materials and those of the U.S. Geological Survey. One should not neglect the progress reports issued annually by the western surveys as well as the annual reports of the USGS. In the latter, each division chief (like Gilbert), particularly in the early years, gave an account of his year's official activities.

I found graphic materials, both photographic and cartographic, especially useful. The National Archives (Still Pictures Section) holds the photographic collections of the Wheeler and Powell surveys, listed under the name of the photographer—W. O. Beaman, Timothy O'Sullivan, William Bell, and Jack Hillers. More useful, however, were the substantial (thirteen volumes) number of photographs that Gilbert took during his tenure with the USGS. These, with captions, are stored in the U.S. Geological Survey Photographic Library in Denver. Nearly all the photographic illustrations in GK's reports and articles are cached here, as well as a number of photographs by Willard Johnson, some of which were taken on shared field expeditions. The library also has a fine collection of photographic portraits of most of the geologists early associated with the Survey. Negatives of the photographs Davis used in his memoir of Gilbert are kept at the National Academy of Sciences Archives, and I found several interesting photographs housed in the Smithsonian Archives (Record Unit 95). My major source for map materials was, of course, the Cartographic Records Section of the National Archives.

A substantial portion of my research consisted of examining the official papers of people who knew Gilbert. At times letters from Gilbert were uncovered; at other times, gossip about Gilbert recorded by those around him; and, for those people for whom I sketched intellectual portraits, necessary insights and documentation. The Library of Congress held several useful collections—the W J McGee Papers, the Samuel Frank Emmons Papers, the George F. Becker Papers, and, especially valuable for its frequent inclusions of Gilbert references, the journals of C. Hart Merriam—the field journals for 1899 and 1903, the home journals for 1900 to 1918. The Arnold Hague Papers are in the National Archives. The William Morris Davis Papers in the Houghton Library at Harvard University contain a number of letters exchanged between Gilbert and Davis. William Holmes' "Random Records of a Lifetime Devoted to Science and Arts, 1846–1929," stored in the National Portrait Gallery, has a number of references to Gilbert and Dutton. The Holmes Papers in the Smithsonian Archives have several useful items, and the Marcus Benjamin Papers, also in the Smithsonian Archives, have biographical material on both Gilbert and Dutton collected in 1886.

Finally, through the help of Michele Aldrich, I have examined the correspondence between Gilbert and Thomas Chamberlin in the Chamberlin Papers, the Special Collections of the University of Chicago Library.

Biographical Memoirs

Andrews, E. C. "Grove Karl Gilbert." Sierra Club *Bulletin* 11 (1920–23): 60–68.

Barrell, Joseph. "Grove Karl Gilbert, An Appreciation." Sierra Club *Bulletin* 10, no. 4 (1919): 397–399.

Chamberlin, T. C. "Grove Karl Gilbert." *Journal of Geology* 26, no. 4 (July–August 1918): 375–376.

Davis, William Morris. *Biographical Memoir of Grove Karl Gilbert, 1843–1918.* National Academy of Sciences Biographical Memoirs 21, no. 2. Washington, D.C.: Government Printing Office, 1926.

———. "Grove Karl Gilbert." *American Journal of Science*, 4th ser. 46, no. 275 (November 1918): 671–681.

DeFord, Ronald. "Grove Karl Gilbert." *Dictionary of Scientific Biography.* New York: Charles Scribner's Sons, 1976.

Fairchild, Henry. "Grove Karl Gilbert." *Proceedings of the Rochester Academy of Sciences* 5 (1919).

———. "Grove Karl Gilbert." *Science*, n.s. 48 (1918): 151–154.

Mendenhall, W. C. "Memorial of Grove Karl Gilbert." Geological Society of America *Bulletin* 31 (1920): 26–64.

Merriam, C. Hart. "Grove Karl Gilbert, the Man." Sierra Club *Bulletin* 10, no. 4 (1919): 391–396.

Merrill, George. "Grove Karl Gilbert." *Dictionary of American Biography.* New York: Charles Scribner's Sons, 1927.

Pyne, Stephen J. "Grove Karl Gilbert: A Biography of American Geology." Ph.D. dissertation, University of Texas at Austin, 1976.

Smith, George Otis. "Grove Karl Gilbert." U.S. Geological Survey *Annual Report* 39: 11. Washington, D.C.: Government Printing Office, 1918.

Appended to the W. C. Mendenhall memoir is a comprehensive bibliography compiled by B. D. Wood and G. B. Cottle. Only the Basin Range study (which was published posthumously in 1928) is lacking.

Index